WHY CAN'T YOU TICKLE YOURSELF?

WHY CAN'T YOU TICKLE YOURSELF?

And Other Bodily Curiosities

INGRID JOHNSON

WARNER BOOKS

A Time Warner Company

This book is not intended as a substitute for medical advice. The reader should regularly consult a physician in matters relating to his or her health and particularly with respect to any symptoms that may require diagnosis or medical attention.

Copyright © 1993 by Jeanne K. Hanson
All rights reserved.

Warner Books, Inc., 1271 Avenue of the Americas, New York, NY 10020

W A Time Warner Company

Printed in the United States of America
First Warner Books Printing: September 1993
10 9 8 7 6 5 4 3 2 1

Library of Congress Cataloging-in-Publication Data

Johnson, Ingrid.
Why can't you tickle yourself? : and other bodily curiosities /
Ingrid Johnson.
p. cm.
Includes bibliographical references.
ISBN 0-446-39395-9
1. Human physiology—Popular works. 2. Body, Human—Popular works. I. Title.
QP38.J73 1993
612—dc20   93-10192
                                                CIP

Cover design and collage by Cathy Saska
Book design by L. McRee

This book is dedicated to laughing.

Contents

Acknowledgments

Grateful thanks go to Dr. Henry Meeker of Edina, Minnesota, and Dr. Marentette of Minneapolis, Minnesota, for checking the chapters for accuracy.

Introduction

Laughing, sweating, growing hair, itching, urinating, sneezing, yawning, dreaming, maybe even trying to tickle ourselves—all these bodily functions and many more keep us busy, very busy, every day. It even makes you wonder that we can get anything else done. Just take a look at the table of contents, and you'll probably agree.

This book focuses on our fifty most basic bodily functions, the down and dirties, so to speak. You won't be reading here about how the brain's hippocampus affects memory or why we reject organ transplants, but rather about our daily but not dainty burpings, snorings, and hiccoughs.

If you have a human body, you are likely to be interested in "fun facts" about these fifty favorite functions of mine. For example, did you know that laughing is truly aerobic exercise, that coughing forces air up and out of your throat as fast as several hundred miles per hour, that flushing in anger reddens not only your face but the inside of your nose, stomach, and—if you are female—your vagina too? The book will be crowded with information like this, all accurate and, I hope, immensely entertaining.

It is even interesting, I think, to ponder which among this panoply of bodily functions are our human ones alone, and which are not. Dogs blink, cows fart, and cats seem to dream, but probably no other animal can equal us for daydreaming or crying, to say nothing of our range of 7,000 separately distinguishable facial expressions from our eighty facial muscles. Though not all those thousands will be covered, of course, the book will try to get at evolution every so often, by finding us our place in it.

Our bodies are made of atoms and molecules in these proportions: oxygen, 65 percent; carbon, 19 percent; hydrogen, 10 percent; nitrogen, 3 percent; calcium, 2 percent; phosphorus, 1 percent; with significant traces of potassium, sulfur, chlorine, sodium, magnesium, cobalt, copper, fluorine, iodine, iron, manganese, and zinc. This chemistry is wrought into 20 to 30 trillion cells, crowded into each of us. Fine. But wouldn't you rather hear about what it all does—even if it's sometimes a bit inglorious?

Please don't look to this book for basic biology or for medical help. Go to your doctor if you feel sick, and don't hold me responsible for any of your ailments. I'm here to tell you about the fun of rumbling stomachs, hitting your funny bone, hiccoughing, whispering, and much more. Enjoy it!

WHY CAN'T YOU TICKLE YOURSELF?

Backaches

If you have never had a backache, you're not truly human (what you are is another question). We are all just great apes who have evolved to walk upright—and it is very hard on our lower backs. Perhaps we need another boost from evolution to strengthen them. So either go ahead and revert to that horizontal knuckle-walk, or listen to some useful information about backaches. There are the smaller ones, of course, and then the big ones when you'll say "something snapped" and it will hurt for days or weeks.

Your relevant body parts here are the spine, disks, and muscles and ligaments of the lower back. The spine has twenty-four separated bones called the vertebrae and nine fused bones at the bottom to make the tailbone (remember what I said about your early ancestors). The spinal cord, with all its nerves, runs through these bones. The disks are made of collagen and act as cushions between the non-fused vertebrae; they can rupture (not "slip") and so press on the pain receptors in the spinal cord's nerves. The muscles and ligaments need to be strong to hold up the lower spine, and many people get backaches because they sit around for most of every day or are overweight, then do something strenuous all of a sudden that causes their back muscles to spasm. Doctors say that this scenario accounts for most backaches, and, if this is you, get a book on back exercises to strengthen those muscles, and lose some weight too.

The rest of our all-too-human backaches are usually caused by poor posture, poor mattresses, and wrong lifting. To improve your posture, keep in mind that you shouldn't ever stand in such a way as to overdo the curve in your lower back—by slumping *or* by throwing your shoulders back too far. Holding your stomach in and your fanny in, but not locking your knees, will lead to the position best for your back. Sports that require twisting, like tennis and downhill skiing, are particularly hard on it. Sitting is even harder on your back than standing, so get up and move regularly (yes, you can do desk work while standing straight up). And remember that the pelvis should be tilted back, not forward, when seated. This is easiest to do by elevating your feet. When you sleep, avoid squishy mattresses (or sleeping on very hard surfaces) and don't sleep on your stomach, which exaggerates the back's curve. When you must bend down to lift things—even tiny, light things—use your legs and knees and move slowly, instead of turning yourself into a right angle by bending at the waist. One group of back researchers says it is actually easier to bench press 100 pounds (done while lying on your back) than to lean over and pick up a 20-pound bag of groceries from the trunk of a car. And, in general, the stronger your abdominal muscles, the less the need for your lower-back muscles to try to do everything (and sometimes fail).

Backaches lead to about 20 million doctor visits a year in America. If we walked on our knuckles like the chimpanzees, we wouldn't need to go.

Blinking

It may well lack the whimsy of a wink, but a blink is essential to the health of the eye. About once every six seconds, on average, the eyelid quickly closes, spreading a lubricating, cleaning, and moistening secretion over the cornea, then flickers open again. This healthful, watery substance, which includes salts, some mucus, and an antibacterial enzyme called lysozyme, flows from twenty to thirty tiny oil glands along the edge of the eyelid. And I mean flows—your eyes produce enough of this fluid to coat the bottom of a beer can every day.

Normal blinking does not interfere with vision since it happens so fast. Your eyelid muscles barely flicker the light, making them the fastest windshield wipers of all. Over a lifetime, every person busily blinks about a quarter of a billion times.

There are also some slightly strange differences in rates of blinking. People with lighter-colored eyes seem to blink more often, probably because their lighter irises reflect more light into the interior of the eye. And, according to some preliminary research, blinking may even say something about emotions and physical health. Pleasant feelings seem to be associated with less blinking, and anxiety or pain with more blinking. Odder yet, Parkinson's disease and schizophrenia are being studied to see if they create blinking disorders that might be used for diagnosis.

Blowing Your Nose

In the 1700s, a man whose nose still probably sets the world record was exhibited as a freak in England. It is said that this schnozzer was 7 1/2 inches long. But could he do any more than all of us can with that miraculous appendage?

When not blowing and snorting, the nose is particularly clever. It sniffs about five times a second to identify a given odor, analyzes it through millions of tiny cilia high up in the nose and their up to 1,000 different odor receptors, then transmits the information to the olfactory area of the brain's cortex. Unlike any of our other senses, whose business part lies in the brain, the work of smell seems to be done largely on site—in the nose. This probably reflects the incredible importance of smelling far back in evolution and before creatures had very big brains. Our own noses, though hardly the ruling schnozzers of the animal world, can still distinguish among 10,000 different odor molecules.

But not when you're blowing your nose. If you are one of America's 15 million allergy sufferers who are affected by airborne substances, no now-patented piped-in peppermint perfume at work will work on your job satisfaction, productivity, or even your nose. Your immune system will overreact to perfumes like these, also to dust, pollen, and other things, making a mess inside your nose. (See also the section "Sneezing.")

Infinitesimal enemies of a different type make us blow our noses from colds. These are virus particles, very much alive inside our noses during the infection. And, at the end of the cold, that extra mucus is there to help us snort out the now-dead virus particles.

Efforts are now under way to develop a very sophisticated electronic nose. It is being designed to do anything from distinguishing pollutants in the air of the workplace, to testing beers, to perhaps even sniffing out enemy aircraft. It's a good bet, though, that this nose on a chip won't ever have to be fixed up with a box of Kleenex tissues.

Blushing

The human face is subtle and rich in its expressions, perhaps beyond anything in nature. The smile of a man across a restaurant table at the moment you both realize you are in love. The contorted red face of a newborn baby crying hard in frustration. The time you laughed so hard at your husband's joking that your face felt tired and your stomach hurt. The glare of a colleague when you were late to an important meeting. The frown of your elementary school math teacher. The startled look on your high school friends' faces when you set off a firecracker. The deep flush of anger before your father spanked you. The sneer you thought you could see, even over the phone, when she turned you down for the prom. The blush of a lover, one you thought you never would see. Our lives are made of these faces.

Of all of this wonderful panoply, the blush may be the facial expression that is the least understood. Mark Twain put it this way, "Man is the only animal that blushes. Or needs to." But why? We seem to blush from embarrassment, surprise, fear, shame, anger, anxiety, and in flirtation, quite a range in emotional arousal. Yet some people never seem to blush, though they may well have the same physiological response without showing it. Young children scarcely ever blush. And women blush more than men. The only thing researchers seem to agree on is that a blush is part of the nervous system's response to stress.

If the emotional stress is from a very small flash of anger, or fear, then the blush seems to be engineered by the hypothal-

amus. This small, deep-brain collection of brain neurons pre-
pares the body for mobilization; it directs the "fight or flight"
response. And quick outbursts of both aggression and euphoria
have indeed been traced to the hypothalamus, which is in
charge of these primitive emotions. In the case of the blush,
it would be the briefest of mobilizations, a surge in the blood
vessels of the face that dilates them fast, then, almost just as
fast, fades.

If, instead, the emotional stress that makes us blush is the
kind less essential for survival—perhaps even something we
partly learn to do—then it is under the direction of the body's
limbic system. Injuries to the brain's hippocampus and amyg-
dalaareas—the extensive limbic system's directors—have indeed been shown to make people seesaw quickly in their
emotions. And yet other blush theories point to the frontal
lobes of the brain as involved too, since they help to control
some of our emotions.

In other words, the blush is still somewhat of a mystery.
But a beautiful one. As John Donne wrote in a lovely seven-
teenth-century poem, "Her pure, and eloquent blood/Spoke
in her cheeks, and so distinctly wrought,/That one might almost say, her body thought."

Body Odor

Of all the bodily functions in this book—each one common and "down and dirty"—this may be the one you are least likely to admit to. But everyone stinks sometimes, and we certainly have a healthy fear of it. Otherwise, why would Americans spend more than $1.6 billion on deodorants every year? Of course, some of our natural smells are even quite fetching, so read on.

Underarm perspiration contains more than 100 chemicals, not all stinky. The stinkiest one, called three-methyl-two-hexanoic acid, is produced by bacteria that live in your armpit. As these bacteria metabolize, or eat, some of the chemicals coming out of your sweat glands there, you stink. The degree to which you stink is largely dependent on whether you harbor a lot of these bacteria. (They can be rearranged in the shower, but they don't leave.) Men have more of them than women do. Men also have more steroid hormones in their systems, which is what lends a musky odor to their sweat compared to women's. Your stinkiness is also related to the size and number of the sweat glands in your underarm area. Men have more of them and so do some ethnic groups. Caucasians lie in the middle between Negroes and Asians. Japanese and Korean women have the fewest of these underarm sweat glands of any group.

Beyond our stinky-number profiles, we each have an individual, milder, odor, one that is with us every day of our lives. This is what conveniently allows bloodhounds to find us even when our trail through those wet and windy woods is a week old. These dogs can distinguish any two people except identical twins, who smell noticeably different only if they have eaten very different foods for quite a while. There have been experiments in which people, too, have been able to identify each others' smells; one such study used T-shirts that had been worn by men, by women, and by the individual subject's mate. Then there is that curious phenomenon by which women living together, as in a college dormitory, synchronize their menstrual cycles; researchers say that one of the women emits more of the relevant smells, encouraging the others to match it, gradually and without even realizing it. Some people can even identify others by a whiff of their hands, if challenged to do so; and mothers and babies can recognize each other by their smells from day one. If you think about it, you probably love the signature scent of all the people you love. Our smells are so basic that in some languages the word for "to greet" is the same or similar as the word for "to smell."

Back to the stronger stinks. The little glands in our underarms that can act up are also found in our genital areas, and thereby hangs a tale, you might say. They create the odor of sexual excitement, and this chemistry has been isolated in men. People in one study thought the genital scent was a bit like sandalwood perfume. We are all animals, after all, and it shouldn't seem strange that odors play a role in this area. Our civilized ways have masked their influence, however, to some degree, and for good and bad.

Women can detect stinkers better in general than men or children. Their estrogen hormone is the key. And when estrogen levels are their highest, during ovulation, women are especially good at smelling. Perhaps to detect a man in the woods? Or respond to him better? All very curious.

Perfumes and aftershaves mask our natural scents, for better and worse. Marcel Proust, for one, was preternaturally sensitive to this kind of artifice. He once said to a man who sported merely a perfumed handkerchief, "My dear friend, the last time you were so good as to come and see me . . . I was obliged to take the chair you sat in and keep it out in the courtyard for three days." He is famous for using the delicate smell of madeleines (small, rich pastries) in *Remembrance of Things Past*, but maybe he could have described chair-seat scents instead. Smell, for all of us, evokes memories wonderfully.

Bruising

A bruise is a sign of bleeding under the skin, usually from a dull blow that caused no puncture wound but injured blood vessels. Your body controls this internal bleeding just fine. First, your blood vessels narrow to reduce the flow of blood. Then blood platelets arrive to handle clotting by sticking together to plug the flow. Next, proteins are activated to make this healing barrier permanent. Finally, your blood-born macrophages, the scavenger cells, arrive to carry off the dead cells. This last is what turns the bruise gradually from black and blue (the dead blood cells and veinous blood) to lighter colors as the concentration of detritus thins.

If your bruises don't heal with alacrity, or if you bruise easily, ponder these facts: Aspirin, even one tablet, can slow the blood platelet function for a whole week. Other substances can interfere with it too, such as anti-depressants, anti-inflammatories, some asthma medicines, too much alcohol, and even cough syrups. What helps to strengthen the walls of the blood vessels is enough vitamin C, folic acid, and vitamin B12. And bruises come more easily, they say, if your fat cells are particularly plumped up—there seems to be more to bruise.

Burping

The formal word for this bodily function, fond favorite of kids everywhere, is *eructation*. And some of the gases you emit by mouth when you eructate are flammable. So don't burp out the candles after that fancy dinner or you may cause a fire and damage the tablecloth as well as that candlelit relationship. (Information like this is worth the price of this book, don't you agree?)

Swallowing air will make you burp, and the more stress you feel, from whatever cause, the more air you usually swallow. Chewing gum, drinking carbonated drinks (including beer), and just eating and drinking fast lead to an excess of swallowed air too. In the case of beverages with carbon dioxide (used to make them fizzy), that gas comes right back out of your mouth when your stomach warms up the drink. The human stomach holds about 2 1/2 pints of material, and there is usually not room for too many gulps of air in there with all the food; after all, even foods eaten four to five hours ago are still present in the stomach in one form or another.

Apart from what you take in, your stomach and large intestines make gas too. (See the section "Farting.") The gas made by the stomach, to the tune of a few quarts per day, is a by-product of ordinary food digestion. And it can mix with the food and proceed down to the intestines—or come up and add to that symphony of burps.

Have you ever burped many times in a row, little burps like a volley of cap pistols? If so, you were probably swallowing more air in with every burp out. If you are doing this, you will see your throat seem to gulp as you burp.

Let us end with one simple bit of advice: At least close your mouth when you do that burping . . . belching . . . eructating. And say "Excuse me" the way your mother told you to.

Choking

You are supposed to breathe in through your trachea or windpipe, and swallow through your esophagus, but both open at the back of the throat—and sometimes you are too careless or sick to keep this straight. You breathe in a bit of saliva, or water, or vomit, then choke briefly, then cough it out. Or you scarf down big chunks of meat like a wild animal, or don't chew a hard piece of vegetable and so get a piece caught over your bronchial tubes (near the opening of the windpipe). If you are unable to draw in or expel air along with the object in a cough, you can actually die in just four or five minutes. When the brain is deprived of air, it quickly becomes damaged, and, when it can no longer order the lungs to breathe, the face turns bluish—and soon that's it. You may choke, which is your body trying to start a cough, after all, but you cannot. About 3,000 Americans choke to death every year.

Coughing

In your lifetime, with any luck, you will breathe in about 75 million gallons of air in some 600 million breaths. It is likely that a few tens of thousands of these will be a special type of breathing called coughing. This may seem like one of the more rackingly unpleasant bodily functions, but it is essential to your health. Those who cannot cough—usually people under general anaesthesia—are in danger of harm to their lower respiratory tracts, since they are unable to remove irritations, which coughing does for us very nicely.

The causes of coughs make a long list: an irritant in the air (pollution or smoke, for example), an irritant in the throat (such as a bit of food), a cold virus, an asthmatic condition, a wide range of allergies, a chronic sinus infection or chronic bronchitis, pneumonia, lung cancer, tuberculosis, tonsil or adenoid enlargement, nervousness or another emotional condition, and even heartburn. These nastinesses all add up to cause about 30 million doctor visits in this country every year, courtesy of the simple cough.

The most common cough is the one that comes from the common cold, and what happens during it is typical. The irritant, mucus, is cleared from the respiratory tract by being hurled out in an explosive burst of air. The lungs act as powerful bellows and expel this offender at the speed of several hundred miles per hour. At the same time, the vocal cords close partly, allowing this "wind storm" to rush past the larynx, creating the distinctive noise.

All very impressive. But if it keeps happening all night

long, breaking your sleep, or if you cough so hard that it hurts or even fractures ribs and ruptures stomach muscles, you may fall for one of the "snake oils" in the drugstore. And who could blame you, even though these cough medicines don't help much, if at all. With an ordinary cough, you are far better off coughing up the mucus and getting rid of it. Your cough is just trying to help you.

Crying

If you start sobbing while strolling the jungles of New Guinea, the people there will understand. If a tear falls while you are talking to a couple alongside their pile of seal furs and kayak in Siberia, they too will know your emotions. What these people will understand is not necessarily why you are sad but simply that you are, for crying indicates an emotion that centers on sadness. Every cross-cultural study ever done on the basic human facial expressions has found that they are universally understood, universally human.

This commonality extends backward in time too: By 3,000 B.C. the ancient Babylonians had written a tome, called the Twitching Books, that was all about the expressions of the human face—and they are also our own. Who knows if even the Neanderthals cried, the way we might, over sad stories around the campfire?

Crying is complex, involving not only a spectrum of sadnesses, but beyond that anger and frustration, excitement, awe or spiritual uplift, shame and embarrassment, even fatigue and happiness. In expressing these feelings, our face crumples up in helpless grief, or tightens a bit more in the fierce crying of frustration, or stretches somehow—eyes wide—as just a few tears fall in sadness. The throat may feel a lump or be convulsed with sobbing. And our eyes show our different feelings when they make tears, a few or a deluge. Even people who think they don't or can't cry may well release a few emotional tears when sad, ones that remain within the rims of their eyes.

Tears are of two kinds, emotional and non-emotional.

17

Both are controlled by the lacrimal glands, two main ones and sixty more accessory glands tucked mostly inside the linings of our upper eyelids. We have two types of these glands and each secretes a different kind of fluid, one an oil, and the other water, with extra eye cells contributing mucus. When the tears are non-emotional—designed to clean the eyes routinely or to remove an irritant—the eye's system of ducts drains most of the combined fluid into the nose and throat. (It is because these ducts work less hard at night that we often wake up with a bit of "sleep" in the corner of each eye.) When the tears are emotional ones, they can spill over. And these tears are much more complicated.

Scientists who have studied emotional crying emphasize this excretory function. In crying, as in sweating, exhaling, urinating, and some other bodily functions, a substance is made to leave the body. But why? If we need to express feelings of sadness, or to signal for help, why don't we just grimace or howl or something—instead of having developed a whole new miniature excretion system? The reason is that tears seem to actually relieve the stress we are experiencing. In studies along these lines, volunteers have been shown sad movies in a lab, and then been asked to cry into tiny test tubes. In their tears were found more than just the proteins that control infection and regulate the eye's chemistry. These emotional tears contained certain hormones and neurotransmitters, notably prolactin, adrenocorticotropic hormone, and leucine-enkephalin. The first substance is associated with both milk production and stress. The second is the hormone most associated with stress. And the last is a known pain-reliever created by the brain to modulate pain, moods, and the immune system's response to stress. There are also, in lesser concentrations in these tears, other neurotransmitters related to the brain's behavior under stress. Crying tears appears to be good for you.

* * *

Surveys of normal, healthy adults aged 18 to 75 came up with this: Women's crying diaries revealed up to twenty-nine "cries" a month, but men's only up to seven. The few people who suffer from a disease called familial dysautonomia—which means they cannot cry at all—react severely to mild stress. Instead of producing tears, they sweat profusely, drool, and break out into skin splotches. The medieval physiologists who thought that tears rid the body of "bad humors" were not so ludicrously off the mark.

Moving beyond our own species for a moment, we find a lot of arguments over whether animals cry, and whether the eye moistenings that are indeed seen in some of them are non-emotional eye washings or truly emotional tears. Bull elephants wet their cheeks when in musth (the male sexual state), but this seems to be an arousal signal. Quite a few seabirds have damp eyes sometimes, but this is almost surely related to a system that clears salt from their eyes after ocean dives. The tears of sea otters and seals may be similar correctives when they "cry." The great apes do not cry, even when in visible agony over a death of someone close. But more research needs to be done in this area.

People who never or almost never cry might ponder this: It may be a lot worse to turn sadness into anger and to swallow your stress. Go see a sad movie or sit in a quiet, partly dark room with sad music and a sad book. Listen to Shakespeare's King Richard II, who, late in the play said, "You may my glories and my state depose, But not my griefs; still am I king of those." And try to cry!

Dandruff

Amid the forest of hair strands on your head—each about .008 of a millimeter thick and covered with tiny "roofing tiles" of cuticle—may be lurking the dreaded dandruff, the stuff of all those television ads that make us fraught with fear. Actually, every person's scalp must shed its dead cells somehow, and you have about 19 million skin cells in every square inch of your skin, including your scalp. In fact—get ready for this one—during your lifetime you will divest a grand total of about 105 pounds of dead skin from the surface of your body, all leaving in tiny flakes. Fortunately this won't all dump off your scalp on the one day you're wearing your best black.

In other words, everybody has a tiny bit of dandruff. About 60 percent of us have more of it, though still a mild case. And the rest, at least occasionally, have large, oily scales that stick to the scalp for quite a while, then fall off.

Want to have less? Then shampoo more often than you do now, even with your regular shampoo. If the little snowstorms still don't go away (even though it may not be winter), you may be a person who has a naturally very scaly scalp. This scaliness can come from an overload of bacteria up there, from hormones, from a malfunction in the tiny oil glands that decorate the skin, or from not enough rinsing or too much conditioning after your shampoos. Try dabbing very diluted lemon juice or witch hazel on your scalp before a thorough shampoo and rinse. If nothing helps, ask your doctor if something is wrong.

You could also try one of those dandruff shampoos that

are so avidly advertised and that contain powerful chemicals. Unfortunately, your scalp skin will build up resistance to a particular brand after two to three months of use. And then you will have to change to another brand with a different powerful formula.

Daydreaming

We all do it—to entertain our-
selves, to work on our worries, to feel sexually aroused, to
motivate ourselves, to jog our memories, to run over our personal schedules and agendas, to be stars in our own shows, to
fill the spare minute or two while we are waiting at a stoplight.
All these moments add up. People spend more than a third of
their waking hours in daydreaming. We make scenes—or let
them bubble up—that last anywhere from a fleeting thought
to a few seconds, even up to a minute or two at a time.

Daydreams are eerily similar to nighttime dreams in some
ways. Both peak in about ninety-minute cycles, though daydreaming occupies more of our day than night dreaming does
of our night. The content is usually less weird in the dreams
that occur when we are awake, but then that depends on whose
daydreams we are talking about on what day.

Daydreaming is not always a neurotic time-waster either.
It has been harnessed to our almighty efforts at self-improve-
ment, since experts urge us to imagine ourselves victorious in
tennis, to tell ourselves how wonderful we were at that meeting, and to start fantasizing about a different man or woman
(not the one that got away). This is not just smarmy self-help.
Daydreaming can indeed be helpful in solving problems and
boosting our chances for success. And daydreaming about
strangling that obnoxious gum-chewer at work is certifiably
better than actually doing it. It can surely be more creative,
too, to let your own ideas and images percolate for a while
than to stare at somebody else's imaginings on TV.

Other benefits of daydreams range from the more obvious

(they relax us) to the less obvious (they can lower blood pressure sometimes). If recurrent they can even suggest a change of course or career, and they seem to actually rest the brain by switching it to another mode. So daydream on.

Defecating

Via the intestines, it is a long way from your mouth to the other end: about 30 feet. It's also about fifteen hours from food into the mouth to food out that other end.

Here's how the process works in between. You eat food, which means your teeth and saliva turn it into what is called a bolus. (See also the section "Spitting.") This blob then takes a few minutes to be moved by muscles down the esophagus to your stomach. There, gastric juices pour out of the mucus-protected stomach walls, turning the bolus into a slurry called chyme. This takes three to five hours, and you wouldn't recognize your food by the end of that time.

Once slurried, the stuff moves into your small intestine, so named not because it is short—it is a full 22 feet long—but because it is relatively narrow, just about 1 1/2 inches wide. Here, the pitted and twisted walls contract at the rate of about three times per minute, and the nutrients in your long-changed food are absorbed into the blood and lymph streams through its walls. Some of the nutrition is absorbed in the early feet of this progress, more by the end. This intestine does its work in about four and a half hours, usually moving the chyme along at about an inch per minute.

What is left moves into your large intestine (the colon), just about 6 feet long but wider than the small intestine, at 2 1/2 inches across. This organ's job is to absorb the remaining nutrients and to remove much of the water from what's left of your meal. It takes between five and twenty-five hours to remove enough of the water from enough of the food along the way—this water becomes your urine—to make you want to defecate. The waste products sit in your rectum, which is the last 6 to 8 inches of the large intestine, until you do. Then the large intestine contracts, the sphincter muscle in your anus relaxes, and so do you.

When the large intestine does not remove enough of the water, you get diarrhea. But whatever its form, the feces contain some water, as well as indigestible materials such as cellulose (from bran, fruits, vegetables, and so on), dead mucus cells (from the 100 or more square feet of surface of the digestive tract), and billions of bacteria.

The most common intestinal disorder is irritable bowel syndrome—about 15 percent of Americans have it. In this condition, constipation and diarrhea alternate, and abdominal cramps and bloating and burping can be common. It seems to occur when the muscles in the intestinal wall contract too much. Causes include infections, coffee, alcohol, and stress, even, apparently, the weather and spicy food can make it worse. A change of diet helps—add more fruits and vegetables but avoid those that cause the most gas (see the section "Farting"). Also investigate the possibility that you have lactose intolerance, the inability to properly digest milk products. If you have an irritable bowel occasionally, don't worry about it.

More serious is Crohn's disease, a chronic inflammation of part of your digestive tract. You can distinguish it from the thoroughly unpleasant but harmless irritable bowel syndrome because this disease often includes pain on your lower right side shortly after eating and also a low fever. If this is you, go to the doctor.

Another potentially serious problem is diverticulosis, a condition in which little pouches of tissue protruding out of the intestinal wall become inflamed. As you age, your chances of experiencing this illness are very good, especially if you have been a lifelong fan of laxatives and low-fiber foods. Go to the doctor if you suspect it.

Then there is perhaps the most serious bowel disorder: ulcerative colitis. If you have a fever, bloody diarrhea, pain, blockage, and loss of weight, go immediately to the doctor. Without treatment, your liver and kidneys can become damaged, and cancer of the colon or rectum can follow.

I hate to end on this note. After all, a visit to the bathroom usually turns out fine.

Dizziness

Twirl yourself in a tight circle, ride a rollercoaster upside down, take a trip on a spaceship, look over the edge of a steep cliff, or watch a film where the scenery flies by fast—and you'll feel dizzy, probably happily so. Out of these pleasant excesses, after all, whole industries have been born.

But ponder ordinary, daily dizziness as a physical phenomenon for only a moment and you will realize how dangerous it can be. Without a well-honed sense of balance, you could easily fall down and hit your head or break a bone. In the early days of our species a dizzy person could fall at the hunt for food or even get eaten by a predator. Even today, dizziness can interfere with body orientation when running or leaning over or even walking up or down hill. Dizziness is something all animal bodies, even the very primitive bodies of ancient creatures, have evolved to protect against. We need a hardy sense of balance, one that we can easily recover in a very whirling world.

Our protective adjustment against dizziness is centered in a very small, inconspicuous area of our bodies, the three semicircular canals; these are tucked near the cochlea in the inner ear. One of these canals is sensitive to our backward and forward motion, another to right and left movement, and the third to up and down moves. The fluid in this canal system, called the bony labyrinth, sloshes a bit whenever we move our head; this movement is relayed to the brain, which then automatically sends out messages to adjust our position. The three canals lie at right angles to each other, and at least one

27

of them picks up any motion. Dizziness is really just the warning they provide so that the brain can restore our balance before we fall down.

Part of these semicircular canals are tiny chambers with even tinier hairs that also contribute to balance. They detect not the motion but the position of the head vis-à-vis gravity or any other accelerating force. Again, they work through connected nerve cells to alert the brain, which then sends out orders for the body, including the eyes, to compensate for any unusual orientation. Notice, the eyes of a dizzy person move back and forth fast, trying to keep things in view, while the dizziness has the scene scrambled.

Dizziness is caused not only by movement. Heat, for example, can make us dizzy, by changing the density of the fluids in the semicircular canals or by changing the flow of blood to the brain. Diseases can create dizziness, too. For example, in Ménières disease the ear permanently develops excess fluids in the semicircular canals, making for vertigo so severe at times that it creates vomiting and sweating; the worst cases require surgery, to drain the fluid.

Getting dizzy every so often and for just a while is normal, of course. If only there had been amusement parks during the days when we were all cave people, we might have already evolved to compensate for their sloshing shenanigans. But we'd be having less fun.

Dreaming

Probably for as long as dreams have been spun out of the human mind, like silk threads from a spinning wheel, people have been wondering about dreaming and the meanings of dreams. They have been considered to be many things: The landscape of the soul revealed. A physiological flurry to clear the brain's synapses, or replenish its communications chemicals. An order from God, or the devil. A daily chance at ego enhancement. A foretelling of the future. The deep structure of the brain in problem solving and creativity. A time to store memories. A reason for nightly erections and lubrications. A safe daily insanity. A visit to the world of the dead. A resifting of the day's learning. A guide to the unconscious. And probably even more.

Records of dreams and their interpretations have been found etched into clay tablets from as long ago as 5000 B.C., and that was probably not the beginning of this deeply human curiosity. We will never know whether people have been dreaming since there were people on earth, but we do know that the dreams of children before the age of eight or nine contain only a few scenes without a script. After that, they start spinning yarns like the rest of us, most likely shaping the nightly flux of the neurons with their own fears and wishes.

Some basic facts: Adults dream at least one and a half to two hours a night, spread over several dreaming periods, whether they remember it or not. If we are deprived of dreaming by sleeplessness, or by drugs such as alcohol or barbiturates, we may become subtly restless but not psychologically disturbed. We will, however, dream longer later, as if to compensate. We have evolved to render unconscious our 10 billion brain cells during sleep, then use them peculiarly during dreaming. Beyond this, much mystery remains, though one suggestive study on rats discovered that when the animals were prevented from dreaming for ten days, they began to eat like pigs yet died of starvation during their waking hours. Almost all dreams come during the unique sleep stage called Rapid Eye Movement (REM) sleep. Under normal circumstances, we enter this stage at about ninety-minute intervals throughout the night, and the four or five periods we spend in it—and so in dreaming—gradually lengthen as the night progresses. Short dreams of a few minutes early in a period of sleep stretch to perhaps an hour's episode near waking. Longer sleepers dream somewhat more than short sleepers, though not commensurately so. Babies seem to dream to some degree at least twice as long as adults do.

While in this REM sleep, we are hardly blobs in the bed. Our eyes dart behind our eyelids, our hearts pound, our breath comes quickly, and our brains remain almost frantically active. Our muscles may twitch occasionally too, but the body is essentially paralyzed, in what may be a safety measure that has evolved to keep us from acting out what we are dreaming when our senses are not alert enough to keep us safe. This strange paralysis, found only during REM sleep, seems to be controlled by an area within the brain stem called the pons, a very primitive structure. (This paralysis has, perhaps, an echo in our waking state, as we hesitate for a paralyzed second when in a panic.)

The scripts of dreams vary wildly, of course. But a normal person has about three or four times as many anxious, angry, or sad dreams as happy and exhilarating ones, and within dreams, hostile acts are about twice as common as friendly ones. True nightmares are fairly rare among adults, though, and usually occur when one suddenly gets extra REM sleep. Nor is it true that all dreams involve sex; in fact, male erections and female lubrications during sleep are not caused by erotic dream content but involve a pooling of the blood under the control of the primitive brain; they often last for the entire REM period. Dream content does change somewhat with age. Younger adults are said to dream more about moral conflicts, middle-aged people about sex and aggression, and older people about sickness and death.

The content of dreams can be affected by external events too—a real telephone's ringing can be woven into a dream, for example. In one study, dreamers who were splashed with water and then assaulted with loud noises and light incorporated these events into their ongoing dreams, about one-quarter to one-third of the time. Another study showed that merely repeating a word a few times to a dreamer will allow it to be incorporated about a third of the time, provided the word is important to the sleeping subject.

Dreaming can also be brought partly under conscious control, in a process called lucid dreaming. Concentrate hard on a subject or a feeling as you fall asleep and, with practice, you may be able to create a dream around it. Much puzzling over a difficult and specific problem has even led, in several famous cases, to people dreaming the answers. The German scientist Friedrich Kekule dreamed of snakes twisted in a circle, then woke to realize that this was a major advance in organic chemistry—the structure of the benzene molecule that he had been trying to discover. A professor of the ancient Assyrian language dreamed his key to the translation of Nebuchadnezzar's stone. Musicians from the early hymn-maker Caedmon down to Wagner have heard new music in their dreams, and Mary Shelley thereby got her vision of Frankenstein. Some one-third of college students in one dreaming study reported that they had solved problems of various kinds in their dreams.

There are a lot of myths about dreaming. Some people swear that they never dream, yet need only to set alarms for various times of the night to surprise themselves in one of their many unremembered dreams. (A notepad by the bed enables them to preserve it until morning.) Early risers are especially prone to forget their dreams, and so believe that they never happen. Dreams seem to be stored only in the short-term memory and cannot be transferred into long-term memory storage unless they are somehow articulated. It seems sad that dreams are so ephemeral, but if they were indeed all stored in our memories, we could easily confuse their content with our waking knowledge, and this could really mess up our lives. Your husband or wife wasn't really unfaithful—that was only in a dream last night.

Contrary to another myth, dreams are indeed always in living color, unless the dreamer has been blind since birth and so has never experienced visual imagery of any kind. And, because the waves of nervous activity that launch us into our dreams activate the parts of the brain that produce vision, vision is their main sensory mode, though the sense of touch and movement are also always engaged to some degree. Sound is not always so vivid, but about 60 percent of all dreams involve some hearing. Taste and smells are present much less often. Blind people who have had sight earlier in life do see in their dreams, though they gradually do lose most visual images and come to dream in the sense they use the most.

And, contrary to some people's opinions, dreams are not tightly associated with higher brain function—after all, the two animals that dream the most are the opossum and the ferret. And neither are dreams experienced instantaneously. They occur in the same kinds of "real time" that plays and films do. Dream on.

Now let us end this bodily function with perhaps my favorite passage about dreaming, from Shakespeare's *The Tempest*:

> *Our revels now are ended. These our actors,*
> *As I foretold you, were all spirits, and*

Are melted into air, into thin air;
And, like the baseless fabric of this vision,
The cloud-capp'd towers, the gorgeous palaces,
The solemn temples, the great globe itself,
Yea, all which it inherit, shall dissolve;
And, like this insubstantial pageant faded,
Leave not a rack behind. We are such stuff,
As dreams are made on, and our little life
Is rounded with a sleep.

Ears Popping

This little popping—which sometimes sounds like a whole volley of tiny cannons firing off one by one—is your ear adjusting to a difference in air pressure that has developed between outer ear and middle ear. The pressure equalizes when the air at the two places and pressures mingle; this happens when you unconsciously open your eustachian tubes, which connect throat to middle ear and which are usually closed. When the tubes have done their jobs, the membrane of the eardrum snaps back into place, with its characteristic auditory flourish.

The problem that promotes the pop occurs most often and most dramatically when you are in an airplane that is taking off or landing. Even though flight cabins are pressurized, there is still a difference between the heavier air pressure at ground level and that higher up, where the air molecules are less densely packed. On the way up, there will always be a bit of that "thick," ground-level air stuffed into your middle ear, while the outer ear has just been flooded with new, "lighter" air as the plane ascends. During landing, the situation is reversed. It feels a bit more uncomfortable on take-off than on landing because on the way up the tympanic membrane (eardrum) is bowed out farther toward the outer ear, pushing on more pain receptors there. Pop pop. Relief.

Farting

Farting is caused in its greatest quantities by foods that are also very good for us, the most celebrated being beans. The four worst for gas are probably soybeans, pink beans, black beans, and pinto beans, but all beans have two carbohydrates that are not absorbed well in the intestines. (These culprits are stachyose and raffinose.) What cannot be absorbed just lies there and is gradually fermented by the friendly bacteria that live in the colon; it is this fermentation that creates the offending gas. The same situation pertains, to a lesser extent, to many other foods, commonly broccoli, cauliflower, and brussel sprouts (the cruciferous vegetables that seem to protect against some cancers); spinach and eggplant; and large helpings of fruit and bran. Incidentally, the fermentation potential of fruits is the process behind wine as well as flatulence.

These foods—and individuals can have plenty of unique sensitivities to others too—all have their own odors, and once fermented, they smell even worse; this is what lends "perfume" to the main gases already in your bowels, which would ordinarily be close to odorless. Then out it comes. The composition of an average such explosion is about 59 percent nitrogen, 21 percent hydrogen, 9 percent carbon dioxide, 7 percent methane, 4 percent oxygen, and sometimes a trace of hydrogen sulphide (that "rotten egg" smell). A look at this list reveals that every fart is also a contribution to the greenhouse effect. A more careful look indicates that farts can even be flammable (if enough hydrogen and methane combust). Of course, farting has to happen to everyone sometime, no matter what you eat, or what your environmental politics, or whether you have a fire extinguisher at hand.

Sometimes excess intestinal gas makes people think they have worse conditions, such as an ulcer, angina, or a gallbladder problem. Alternatively, sometimes gassiness is indeed a sign that something is wrong. It can be a marker for intestinal disorders such as lactose deficiency (an inability to absorb the sugars in milk products), irritable bowel syndrome, diverticulosis, or inflammatory bowel disease. But usually gas is just its own unpleasant self.

Food is not the only thing that produces farting. Eating in a hurry or when under stress, swallowing a lot of extra air, sucking on candy or cigarettes or gum—all these can contribute to it. Even sex has been implicated in the origin of flatulence; during any variety of sex, the rectal muscles can become relaxed, opening the door for the nasty gas.

All this is actually better, though noisier, than not being able to release any gas at all. People who will not or cannot move around sometimes build up too much trapped gas in their abdomens. Fluids, exercise, and some antacids can help to bring these people back to the happy company of farters.

One more bit of good news: Some scientists are poised to begin genetic engineering on beans. They want to keep their good points but do away with much of the gas of this "musical fruit."

Fingernail (and Toenail) Growth

Are you between twenty and forty years old? Or pregnant? If so, lucky you: Your fingernails and toenails are growing faster than the rest of the population's. On anyone's extremities, though, nails grow more in summer than in winter, and less when you are ill. And all of our fingernails grow about three times faster than our toenails. The average rate of growth for a fingernail is one inch in eight months—though people vary and nails vary too—and the middle fingernail grows the fastest, the thumb and little finger the slowest. Usually a fingernail will end up breaking by the time it reaches five inches long, and a toenail by the time it is almost that long, though there are people who have grown theirs much longer by just sitting around and not using their hands or feet very much.

A nail is a plate of dead cells made of a tough protein called keratin. Your hair and the outer layer of your skin, all made of thoroughly dead cells too—are also keratin. These dead nail cells are continuously being pushed out by living cells in the nailbed. This little growth factory of blood vessels is mostly hidden under the bit of skin at the base of the nail, but the white half-moon effect is part of the operation.

Fingernails and toenails look hard but are actually porous. Water can move through them about a hundred times quicker than through your skin. This makes for potential damage from nail polishes and polish removers, which dry the nails. If you use these products, you should let your nails "breathe," free of polish, for two to three days out of every month, otherwise, they will yellow. Other things that dry out the nails include very hot water and various household cleaning products. Hand cream is good, but quaffing gelatin drinks does not help most people to improve their nails. If your nails change, it could be a sign of any number of diseases, so visit your doctor.

Flushing in Anger

A beet-red face, the spiking up of blood pressure and heart rate, dilated pupils, even flared nostrils and the erection of the body's hair follicles (to make your "hair stand on end")—all are the outward and visible signs of the inward maelstrom of anger. A flushed face, the most obvious of these, is also accompanied by a flush in the membranes of the stomach lining, nose, and vagina. Like humans, apes and monkeys also flush in their own rages, tantrums, blusters, bridles, and fumes.

The flush and its anger is masterminded by the hypothalamus, that king chemist of the body's emotions. Injury to it causes odd anger reactions. It coordinates the work of the two divisions of the body's nervous system—the sympathetic and the parasympathetic; the former accelerates the heart and constricts the blood vessels, while the latter helps the anger reactions to pass. The adrenal glands are quickly activated in anger too, dumping hormones into the blood and shunting more blood to the muscles and much more. All this bodily busyness is designed to make us ready to either fight or flee, a basic and primitive readiness. And that's fine if fight or flight is required, or if this cascade of reactions does not happen too often.

The problem is that we cannot slug the boss or flee our relationships every time something angers us, and people who have large dollops of anger in their emotional stews end up stewing in them a lot. Their sympathetic nervous systems chronically order too large a dose of stress hormones, such as adrenaline and noradrenaline, and their parasympathetic nervous systems do not provide enough of the hormone acetyl-choline, which is supposed to step in to balance this flood.

As these angry peoples' lives progress, their arteries and hearts are weakened from the effects of excessive mobilization, and even their livers and kidneys, which help to make choles-terol, are bathed in excess fats released in these reactions. The result is a person with heart problems, high cholesterol, and therefore a shorter life. The latest results indicate that chronic anger is more apt to kill than high-fat diets or even smoking. As Shakespeare said in *King Henry the Eighth*, "Heat not a furnace for your foe so hot/That it do singe yourself."

Of course, not everyone whose face occasionally flushes in anger is heading for an early grave. The deadliness seems to be specifically in the hostility and suspiciousness with which some people chronically react to life, and about 20 percent of us fall into this category, researchers now think: Suppressing this rage does not help at all. What helps, though, is to be able to talk yourself out of it in mid-rage ("Why do I need to be so mad at having to wait two minutes? I can read or ponder that problem I've been having on the job.") Or distract yourself by talking to someone calmly, for example, or even counting to twenty. These actions can thin the waterfall of nasty internal reactions that can accompany a flushed face.

Frowning

This unattractive expression is mainly a function of the "corrugator muscle," the one that can move your eyebrows down and closer together and compress even your eyeballs a bit. Charles Darwin, the first scientist to study facial expressions and author of the book *The Expression of the Emotions in Man and Animals*, called the corrugator "the muscle of difficulty" because it is also implicated in other expressions of non-happiness such as glaring and scowling. To the people who think the function of eyebrows is just to sop up sweat, consider that we also need them for our all-important facial expressions. The origin of frowning, in terms of evolution, may well have been in squinting and concentrating, not forehead sweating.

Now, you know what your mommy said: "Don't look like that or your face will stick." She was right though it takes thirty years or so of too-frequent frowning to give you those frown lines from contracting your corrugator muscle too much.

Frowning was also part of probably the most interesting modern study of facial expressions ever conducted. Professional actors were told to make certain "faces" based upon detailed physiological instructions rather than the names of the expressions—for example, "furrow your brow and press your lips together slightly," and so on. Once the actors had made a given face, and just as they were figuring out what it was, their heart rates, skin temperatures, and other relevant signs were taken and found to match the emotion they had unwittingly expressed on their faces. Most surprising of all, when they were asked how they felt, they answered with the emotion whose face-muscle profile they had unwittingly made. Putting on a sad face actually made their heart rates speed up the way sadness can do, for example.

This seemingly "backward" phenomenon is thought to work this way: The facial muscles used in our emotions trigger the autonomic nervous system and thereby the brain's hypothalamus (our mood control center) to begin to "make" the emotions. They do this by slightly altering the blood flow to the brain.

Most people's favorite mood, happiness, was problematic in this study, but if you doubt that facial expressions can make emotions as well as reflect them, just try smiling for a while—you will actually start feeling happier. I know you think I'm crazy and maybe that these scientists are crazy too, but just try it—and with smiling instead of frowning if you don't want to "improve" those frown lines. It actually works. "Let a smile be your umbrella," as your mommy said.

To realize how special our human facial expressions are, we need only to glance back at evolution for a moment. Creatures simpler than the birds have only two "expressions": the mouth poised for attack and the open mouth for eating! But with birds begins a bit of expressiveness; the faces of chickens have actually been shown to help maintain their chickenly "pecking order." Below the mammals in evolutionary complexity, the muscle system that molds the face is much less supple. Among mammals, though, even solitary species such as the bear exhibit some subtlety in facial expressiveness. And the sociable ones, such as the wolf, have developed a significant range of "faces" for use with the pack. Witness the wolf's descendant, the dog, and its "play face," its drawn grin of threat with ears flat back, and its other faces. Not until the anthropoid apes, however, do we see muscles rise to connect richly with movable skin, eyes, ears, noses, and lips, creating a set of facial expressions well beyond the limits of gesture and posture. Apes can flush with anger; chimpanzees smile differently for fright, excitement, threat, and enjoyment. Though the muscle that moves our own ears is indeed deteriorating—you know how few people can wiggle their ears well—we vastly exceed even the chimps in our "faces."

This is all very well, but don't forget to frown less, whether you're doing it from being troubled, hurt, disappointed, or near tears, in pity or anguish or concern, or with a scowl, glower, or sulk. That corrugator muscle just is not good for the look of your face.

"Funny Bone"

Ha-ha. Hitting your "funny bone," or elbow, is not so funny. It makes a tingling, jangling pain that is different than the one that comes when you hit any other part of your body. It hurts so much because the blow pinches a nerve behind the elbow. This "funny nerve"—officially called the ulnar nerve—runs all the way down from the neck to the finger tips and so can radiate pain (one of the jobs of the nervous system) the whole length of the arm. If you hit the bone especially hard, the nerve will swell enough to make the pain last longer.

There is a reason that this bone and this nerve are especially "funny." Unlike most other nerves, the ulnar is not buried in muscle and other soft tissue but lies right next to the elbow bone ready to absorb any hits. And the bone protrudes.

The only other candidate for the amusement epithet is the ankle bone. Though the nerve behind it is better protected (by ligaments) than the ulnar, it is possible to get that unfunny feeling by hitting your ankle bone too.

"Going to Sleep" in Hands and Feet

First your hand or foot feels dead, a flopsy appendage that seems to belong to someone else, the way your cheek feels after the dentist has given you novocain. Then, as it "awakens," there is a curious prickly feeling. All this is courtesy of your circulatory system—heart, arteries, capillaries, and veins—and your autonomic nervous system. They are always working together to maintain the proper blood pressure and to distribute the blood to various parts of the body according to need. And they have "decided" that you don't need as much blood in that particular hand or foot. But why?

Picture yourself inactive, decked out on the couch and not using one of your hands or feet. Also picture yourself allowing the blood vessels to become constricted in that foot or hand, perhaps by sitting on top of your leg or even dozing on top of one hand. When you do this, you are creating pressure and partly blocking circulation in the very smallest filaments of your circulatory system, the capillaries and arterioles, which are well represented in the body extremities such as hands and feet, far from the heart. Nerve endings near these little blood vessels that you are squashing a little then send messages to the brain to release norepinephrine. This neurotransmitter constricts these tiny blood vessels even further, as an efficiency measure. Once the blood supply to hand or foot falls very low, you get that "dead" feeling. Then, when you notice it—and wiggle your foot or hand— the nerve endings get the message from the slightly increased blood flow, and they tell your brain to order the small blood vessels to open and the blood to flow. As full circulation is renewed, you feel a tickle from this action of your foot or hand "waking up."

Hair Growth

Our monkey days may be over (well, for most of us anyway), but every human being still has hair follicles all over his or her body except on the palms of the hand, the soles of the feet, and a few other very small areas. The muscle system attached to these follicles, or tiny tubes, is extensive enough to raise "goose bumps" on our skin or make our "hair stand on end"—both simply involuntary reactions of the follicles to cold or fright. We are a veritable forest of these little hair follicles, about 5 million of them on every human body. At least 100,000 are burrowed into our scalps alone to make our crowning glory (I'm not speaking about the baldies yet).

Why we are still so furry after all this time is a bit of a puzzlement. Once, of course, our monkey hair was needed to shield our skin from the sun, hold in body warmth, and release body heat as well, by serving as wicks for perspiration. But when the first humans left the jungle for the savannah, they had to cope with hotter temperatures and evolved sweat glands that couldn't work properly on skin clogged with hair. Other theories postulate that control of parasites worked better on skin that was not so hairy, encouraging that trend. One might think that when people began to wear clothes regularly, hairiness would be no longer rewarded by evolution at all. But perhaps it was still useful in distinguishing friend from foe in the distance ("Put down your club, Joe, that's Bob with the curly hair"). And maybe hairiness still has the kind of appeal that means hairy people have more children and so pass along their rather hairy genes. The difference in the appeals of hairy

47

men and hairy women may be the reason women's hairiness has become even less since monkey days. Or perhaps it just takes a long time for relatively unnecessary but harmless body features to go away. It might be all of the above! And I for one am grateful that we aren't stuck instead with some remaining feathers or scales, which, after all, are more common than hair across the animal kingdom.

The growth tricks of our personal forests, however diminished, are really rather amazing. The average person grows about seven miles of hair every year from the scalp alone (just picture it all, lain end to end); this growth peaks in spring and summer and, in daily terms, in late morning and late afternoon. If you never ever had your hair cut, its length over a lifetime would be about 26 feet (heavy to carry, definitely). Hair is nearly pure protein, with each strand a compacted string of dead skin cells that have been pushed up and out from the lower part of the hair follicle. The color of hair is created by granules of pigment added on the way up, and, as less of this pigment is deposited, the hair grays.

Each hair follicle on our heads continues its busy work for three to five years, then rests, allowing its product, the hair shaft, to fall out naturally some three to four months later. This is why we lose fifty to a hundred scalp hairs every day, even if none break off for other reasons—and those that do break would have fallen off in a few days anyway, even if no comb or wind blast had touched them. Once rested, the hair follicle then starts its new growth phase. At any given time, about 90 percent of our hair follicles are working, the rest resting. People who think they have especially fast-growing hair actually have follicles with longer growth periods.

A few words about hair growth elsewhere on the body: Our eyebrows never grow as long as our head hair because the follicles there have a growth period of only about ten weeks before they begin to shed. Body hairs also have a short growth period.

And a few about individual differences: Red-haired people have fewer hair follicles overall; and people with curly,
straight, and kinky hair just have differently shaped hair
follicles.

Now for baldness, a phenomenon regulated by sex hormones. In every male, both bald and hirsute, the male hormone furthers the growth of beard and hair throughout life,
but slows the growth of the scalp hair later in life. In females,
the predominating female sex hormone doesn't do much of
either, and so the sexes vary accordingly. The origin of male
"patterned baldness"—the ordinary kind where hair is lost
first at the crown of the head, temples, and forehead hairline—
is heredity acting through the male hormone. To a degree
that is different for each person, the hair follicles simply die.
Women experience baldness tendencies too, as their female
hormones weaken, though it does not wreak any particular
pattern on the head. Their hair simply thins all over as some
of the follicles die, and this does not usually occur until fifteen
to twenty years later than the time men experience it.

One of Shakespeare's characters in *The Comedy of Errors*
has a lovely word of comfort for smooth-headed men: "What
he hath scanted men in hair, he hath given them in wit." If
you are bald, celebrate it!

But to remedy their patterned baldness, men have slath-
ered on almost infinite nostrums. Various foods, vitamins,
and salves have even given way to radiation treatments and
chemical injections on some people's scalps. But the only three
things that seem to work are castration early in life (which cuts
into male hormones and is no bargain!); the blood pressure
drug Monoxidil, in some cases; or a hair transplant, in which
follicles from another part of the body are moved to the head.
The other choice, of course, is deciding that, Samson and
Delilah images aside, bald really can be beautiful.

Besides garden-variety baldness (a bare garden, it might
be said), there is the more serious kind, called symptomatic

baldness. This can be caused by surgery or by typhoid fever, scarlet fever, pneumonia, influenza, or another serious infection; less often, the culprit is a pituitary or thyroid disturbance, a scalp disease, or an injury. This type of baldness is not, however, permanent.

Halitosis (Bad Breath)

The word sounds like a joke conjured out of those million and one mouthwash ads from television, the ones that used to come after the dandruff ads. It may seem especially amusing if it makes your all-too-real teen-age boyfriend keep disappearing and reappearing while smelling of that nasty mouth squirt. And who hasn't secretly breathed into their hand, then checked the smell—and worried, while mouthing that dreadful, comical word *halitosis*?

Having halitosis is not so funny to the person who has it, of course, and, since that might be you—right now, or next Saturday night—let me tell you about some of its causes. The most obvious is eating smelly food. Garlic-laden dinners, beery celebrations, and such can make your breath stink for a while. This kind of halitosis can be masked with breath mints, toothpaste, and other nostrums pretty successfully, and it passes rather quickly too. Metabolic disturbances caused by dieting or disease can also ill-perfume the breath and last longer. A dieter trying one of those uncomfortable, and harmful, no or low-carbohydrate diets will start making ketones in the body, and they smell bad indeed. A diabetic falling into a coma also makes excess ketone. And some kinds of kidney failure can lend the smell of urine to the breath. (This makes your teen-age boyfriend with the acrid mouth squirt sound real sweet, doesn't it?)

Problems in the body that do not occur in the mouth or throat can still create halitosis, and this is not so strange when you think about it. Whatever chemical is produced in the body gets into the bloodstream, since our veins drain all of our body

51

parts. The chemical then moves in the blood to reach the lungs (as all blood does), gets into the air sacs there, and then is partly exhaled in every breath. If the chemical happens to be one that smells bad—presto, you have halitosis.

The above scenario is rather common in people with lactose intolerance. They cannot fully digest milk products and so exhale a bit more hydrogen than they should, which fortunately does not stink a lot. Nastier by far is the similar situation in people who have problems anywhere in their digestive systems; the smells of seriously rotten food or even feces can pervade their breaths. If this is you, or a (former?) friend, go straight to the doctor!

Sometimes even a seriously rotten food smell on the breath is only temporary, caused by a bit of food trapped in the nether reaches of the esophagus. And occasionally there is a breakdown in the bacteria that work on the food in your mouth, lending that smell of old dinner to the rest of what could have been a lovely evening.

Headaches

Some people have headaches only rarely, others have woefully intermittent migraines or cluster headaches, and some have a tension headache a great deal of the time. But no one escapes them entirely, and no one gets much sympathy either for this invisible problem that afflicts 45 million Americans regularly. If our hair stood up on end or something during a headache, it might garner us more kind comments.

The physiological cause of headaches is still not completely understood. The most common theory is that the blood vessels in the scalp—and there are a lot of them, used for carrying blood to your brain and face—first contract and then expand, stretching the artery walls. Pain receptors in and near these arteries register the pain; the brain itself cannot, since it has no pain receptors. A related theory is that muscles in the neck and head contract and then spasm, triggering the pain receptors. Another theory points to the possibility of the wrong levels in the brain of various neurotransmitters. And a last theory argues that the headache begins when a wave of low electricity sweeps across the brain's surface, affecting the local oxygen supply and thereby causing the scalp's blood vessels to dilate. Under all these theories—and they may all be right— people vary a lot in their tendencies to have these things happen to them and so vary in their frequency of headache.

The stimuli for headaches are many and various. People can easily get them from foods and drinks, especially alcohol, onions, wheat, eggs, aged cheese, chocolate, corn, garlic, herring, the monosodium glutamate still found in many Chinese dishes, bacon, drinks with caffeine, and anything to which you might be slightly allergic. The latter includes substances like tobacco smoke, molds and dust, and perfumes. You can also get a headache from eating nothing—a hunger headache. And you can get one from sleeping too much or not enough, from eye strain, bright lights, and loud noise. Viral infections can bring on sinus headaches. Overdoing alcohol causes a hangover headache. Head and neck injuries can result in what is called a traumatic headache. Sex, in some unfortunate people, stimulates an intense sex headache. High blood pressure and premenstrual tension can lead to a headache too. Then, of course, there are meningitis, encephalitis, and brain tumors, though these are very rarely the cause. And simple tension.

In fact, the most common headache is called the tension headache. The pain is a tightening and a pounding sensation, usually in the part of the head that would be covered by a hatband. About a third of sufferers also have some of the symptoms of migraine headaches during their tension headaches, notably nausea and a sensitivity to light.

Usually migraine headaches involve, besides nausea and light sensitivity, a throbbing focus on one side of the head and a sensitivity to noise. There can also be an "aura" (visual disturbance) or a strange mood disturbance preceding this kind of headache. Some 8 million Americans are said to suffer from intermittent, severe migraines.

The other main type, much less common than migraines, is the cluster headache: sharp pains on one side of the head and face that last for twenty minutes to two hours, then repeat themselves one or more times a day for weeks or even months. They then stop for a long time, then start up again.

There are ways to help yourself with your headaches short of surgically tampering with your complex brain. As for tension headaches, some people have had good luck with relaxation techniques and also with biofeedback, enabling them to reduce the dilation of the scalp blood vessels. Contact a headache or pain center. More simply, you can take a hot bath, exercise or massage the neck muscles gently, even use a heating pad on them; in other words, try to stop the original tension any way you can. If you suffer from migraines, take solace from the fact that these decline as you age. Drugs for migraines include the antidepressant amitriptylene (Elavil), beta blockers, and the newest one—sumatriptan. For cluster headaches, both lithium and blasts of oxygen have been used.

Believe it or not, one hospital program found that even severe headaches can be relieved when people laugh hard. That sounds a lot less complicated and at least worth a try.

Heartburn

In the mid-1800s, a man was shot in his stomach at such close range that the bullet made a hole there that wouldn't heal. This was considered to be an excellent opportunity for study by a local doctor, who got the man to swallow pieces of food wrapped in strands of silk. Why silk strands? So the doctor could pull the food pieces out of the stomach hole at various hours after they had been eaten, in order to study the process of digestion, including the gastric juices. This is how it was discovered that hydrochloric acid pours out of the stomach walls into the stomach to help us digest our food. And don't worry about the man: He did fine by wearing a tight bandage over his stomach hole. It kept his dinner inside when it was not being studied.

You're complaining about a little heartburn? Well, go ahead. All that stomach acid and one little valve give 22 million Americans attacks of it regularly. The valve involved is the one at the bottom of the esophagus, the tube that carries food from your mouth to your stomach. It is supposed to stay shut except when allowing the food to go down. But if your valve there is weak—or, more likely, because you ate fast or had a very big meal—it will open slightly even after the meal has dropped well out of sight. This lets some of the acidic stomach juices move part way back up the esophagus, irritating it. The irritation makes for the burn of heartburn, and the opening allows the odors of the food and acid slurry to rise too. The heart is not involved at all—it is just down there in the same general area as the esophagus.

Hiccoughing

Hiccoughing (yes, that's the way *hiccuping* is supposed to be spelled) is a reflex—quite like sneezing, coughing, snoring, vomiting and so on, though serving a much less weighty purpose. This little "whup" is usually irrelevant, something our bodies do automatically in order to remove an irritation and become more comfortable. A hiccough does for us one of three things: help the stomach get rid of a bit of gas; relieve the esophagus of an irritation; or resolve a temporary and (harmless) loss of coordination between two nerves, namely, the phrenic nerves that control the movement of the diaphragm.

It does all this in a kind of tiny "air vomit." The diaphragm tightens a bit, drawing air into the lungs. Then in a kind of spasm, the glottis (which works as a valve) closes as it tries to shut off this air. This turns the air into pockets of noise. Over and over they pop out—the not only mostly irrelevant, but often irreverent, hiccoughs.

It is not funny at all for some people, of course, especially the ones you see on nationwide TV because they have persisted in their hiccoughs. The record sufferer may be a man who has been hiccoughing since 1922. His almost unimaginably rare condition is thought to be caused by severe phrenic nerve irritation. Electrical stimulation or even the cutting of one of the phrenic nerves in the diaphragm is required for such a sufferer to reach quiet and to eat and sleep normally.

A loud "boo" cannot help these hyper-hiccoughers as it sometimes does you and me. Any distraction has, in fact, the potential to help with ordinary hiccoughs, and so does any forced concentration on something else (such as counting slowly or sipping water). But distraction by fear is an excellent resort because it creates an overflow reaction in the sympathetic nervous system (the one associated with emotions); this can somehow swamp the hiccough reaction masterminded by the parasympathetic (or unconscious) nervous system. Perhaps you'll even have a bit of syncopation in the hiccough finale.

Some funny folk cures work too. Make yourself gag, tug on your tongue, gargle water, stroke the carotid artery (the place that pulses in your neck), taste something bitter, massage the roof of your mouth. The probable reason these tricks seem to work is that the short-term hiccough is just that—and was probably going to go away anyway. You simply ate or drank too fast and swallowed air or suddenly got excited.

A somewhat less helpful method for stopping the hiccoughs was offered by Francis Bacon in his scientific writings: "It hath been observed by the Ancients, that sneezing doth cease the hiccough." But, after all, isn't it harder to make yourself sneeze than to make yourself stop hiccoughing? Just let the hiccoughs go away by themselves.

"Ice Cream" Headaches

If you haven't ever had one of these, you will surely consider me crazy. But they are actually quite common. Ice cream headaches—very sharp and very brief stabs of pain—come from eating ice cream, or any very cold food, too fast.

They flash across our consciousness courtesy of the carotid arteries, the big ones that carry blood from the neck to the brain (the ones that make the pulse you can feel on both sides of your neck). When a blast of cold chills the subordinate connections of these arteries in the mouth and throat, they swing into action to restore normal temperature—this is part of their job. To do so, they must carry more warm blood, and that means they must expand. This enlargement, which occurs almost instantly in the forehead, activates pain receptors there. Hence the headache. And hence, too, its fast disappearance—ice cream is no major assault, after all.

Itching

Whether it is from poison ivy, athlete's foot, dry skin in winter, mosquitoes and their ilk, prickly heat or sunburn in summer, psoriasis, dandruff, jock itch, swimmer's itch, allergies, or that old "seven year itch" at home (metaphorically speaking), itching is something no one escapes. We itch somewhere on our bodies just about every day, usually very briefly.

Itching happens when surface skin cells experience an irritant and alert nearby nerve endings (the ones for itching are now thought to be separate from the ones for pain). It is the nerves that make the tingle feeling of the itch. If the stimulus is more serious, histamines are released, causing a true rash. You then itch more.

Scratching an itch is almost a reflex action. The pain from the scratch then can override the itch. Scratching is "contagious"—if you see someone else scratching, you probably will too. And if you think most of your little itches are for no obvious reason, you are right.

In fact, the origin of itching itself is still a bit mysterious. It is useful to know that something is bothering us, of course, and our skin is a very sensitive and therefore useful barrier. But the reaction often seems excessive.

Laughing

People who are still waiting for their dog or cat to laugh at their jokes know how rare laughter is for species other than our human vaudeville. Even laughing hyenas and laughing gulls are not making their noises in amusement. Chimps do seem to have a minimal sense of humor—they titter as a relief of tension, they tickle each other, and even laugh in games of chasing and such—but we humans are the real hee-haws of the planet. We chortle, hoot, guffaw, giggle, cackle, snicker, whoop, titter, chuckle, howl, and belly laugh our way through life (well, some days more than others). Probably more than anything else—since other animals use tools and have sometimes quite elaborate systems of communication—laughter makes us human. Victor Borge put it nicely when he said, "Laughter is the shortest distance between two persons."

Out of the eighty different muscles in the human face, the one that is the true impresario of the laugh is the zygomatic muscle. (See also the section "Smiling.") To make a laugh, it first contracts, thereby congesting blood circulation to the brain—which is why very hard laughing can make the face red—and then it relaxes. This may sound a little scary, but it is actually good for the body, since as the blood pressure rises, then falls, the brain receives a bit of an oxygen bath like that from a very short exercise program.

Laughing is aerobic. Laughing also tenses and then relaxes many other muscles of the body. If you doubt that laughing moves muscles beyond the face, just watch a few people laugh. Even if they don't double over or roll in the aisles, there is a lot of very visible movement connected with a laugh. One eighteenth-century wag said, "It is rare to see in anyone a graceful laughter." But so what!

Much has been claimed in the way of health benefits for laughter, and comedy tapes are now stocked in some hospitals. It is said that laughing releases endorphins (the body's natural painkillers) and cleanses the eyes. The jury is probably still out on a lot of it. But face it: Laughing feels great, and it obviously helps you at least to some degree.

The beginnings of laughter are interesting to contemplate. I've always wondered what our earliest ancestors thought was funny around the old cave, yurt, igloo, or tepee, and we would certainly know more about their thinking and their social lives if we knew their jokes. ("And the antelope that got away, I'm telling you . . .") After all, a laugh is often built on a rather sophisticated notion of contrast between what is and what could be. Unfortunately, all the first choruses of laughter are lost in the mists of pre-history. We do know that the word *laugh* is very old; it comes from the pre-Teutonic *klak*, which experts consider an onomatopoeic word (one, like *hiss*, that sounds like the thing it describes).

Going a little deeper, the purpose of laughing is also thought to be a highly ritualized way of relieving tension. Observe during a party and you may agree: As people meet and begin to talk, they often let out small, almost inaudible laughs; more obviously, they often say jokey things to each other. Also notice that at airports or train stations, where two people who know each other well are meeting after a separation, will often just plain laugh in greeting, which is curious, when you think about it. Bonds among people are also strengthened by laughing, as they laugh triumphantly at some victory over others, or even jeer at outsiders. Even the structure of jokes often involves a building of tension, which is released by the laughter. Notice how people sigh restfully after they laugh hard.

Orgasm

Now I know you want to read more about this bodily function than about something like spitting or having dandruff.

Let's start with a few fun facts to impress that sex partner of yours who is probably reading over your shoulder now—and will, I hope, decide to go out and buy this book too. An orgasm is said to be a muscle relaxant equal to more than ten valiums (and it's a lot safer and more fun). The average sex session consumes about 200 calories per person. Heavy drinking before sex, for men, actually reduces the size of their erection considerably (if they can even get one) and, for women, sometimes interferes with orgasm. Some women are allergic to the orgasms (the semen) of some partners. During sex, your blood pressure rises about 50 percent, which may make sex qualify as excellent aerobic exercise. In a long, sexy lifetime, you might want or need more but will still have fewer than 700 hours of sex. (A lot of this info comes from *The Body Almanac*; see the bibliography.)

Now about the physiology of the orgasm, which is an impressive bunch of electromuscular contractions, spasms really. For men, the climax lasts eight to ten seconds and involves five or six small bursts of semen. For women, the orgasm is somewhat longer, involving three to twelve vaginal contractions in the muscles lying over the network of erectile tissue in the vagina and clitoris. The female climax also involves some uterine contractions much like those in early labor.

Usually this works fine, but many things can occasionally

63

go wrong. Males, obviously, need an erection to have an orgasm, but about 10 percent of them are always impotent and 50 percent are occasionally so. To become erect, areas inside the penis must fill with blood, and both psychological and physiological factors can interfere. Psychological aspects, responsible about two-thirds of the time, include lack of confidence, problems with the relationship with the sex partner, depression, and even wrong information about sex or an uninteresting partner. The physiological reasons include overdoing on alcohol or barbiturates or tranquilizers, having diabetes or certain heart or kidney diseases that can damage the relevant nerves, having an insufficient blood supply, taking some medicines for high blood pressure or other conditions, not having enough testosterone or enough of its precursors, and experiencing problems from surgery. In both sexes, fatigue, from work or sports or lack of sleep, is a primary interference.

Women's problems reaching orgasm have both physiological and psychological causes too. The physical ones include painful muscular spasms around the vagina, a lack of lubrication (making sex painful), and ignorance in her partner about how to make sex satisfying for a woman (for example, ignoring the clitoris). Psychologically, problems in the relationship with the partner head the list, then come distractions such as fear, guilt, a child crying, or the sex partner providing distraction with inappropriate words or techniques. Orgasm is a response that requires concentration, more so in women than in men.

Treatment for orgasmic difficulties: In men, a visit to the doctor to check for organic causes, and, occasionally, even a little pump or "artificial penis." In both sexes, improvements in the relationship and continuing variety in its sex can restore function. Age may slow responses a bit in both sexes, but this should not make anyone worried. People remain capable of orgasm as long as their health holds out. And it's almost never dangerous.

$\mathcal{P}ain$

If we lacked the capacity to feel pain, we would keep leaning on the stove burner until our hands were singed black, not mind that the bus tire rolled over our toes and, and never notice the violent spasms of our hearts until we literally lost our lives. Obviously we need pain desperately, and that is not a masochistic statement. Only about 100 people have ever been discovered born without a complex sensitivity to pain through the nervous system—and they are in constant danger of losing their lives. The effects of pain on the brain are strongly similar to the effects of learning in lower animals, so pain may have been our first teacher. Chronic pain is something else, though.

Pain sensation comes courtesy of the nervous system, an elaborate network in each of us that, if stretched out, would be about 45 miles long. A single nerve cell can be as small as .0025 of an inch long but can communicate with the brain easily. There are about 20 billion of these cells in the body, connected like branches to a tree into the spinal cord and brain, a miraculously effective electrochemical communication system that always works in less than a second.

Many types of nerve cells live in this system, including our pain receptors, called free nerve endings. They signal for us three different types of acute pain: pricking, burning, and aching. These alarm sounds tell us to stop doing things like over-exercising, or taking too hot a shower, or lying down in a blackberry patch.

Of the kinds of pain, pricking pain is the one people usually think of first. Pain receptors are concentrated in the skin, more thickly in the neck and eyelids and less densely on the soles of the feet and tips of the thumbs. They ignore light touch but react very fast to the beginning of tissue damage. They send the message of that dangerous poking to the brain at up to nearly 100 feet per second.

Burning sensations travel more slowly, at less than 10 feet per second, which is why a pricking pain is typically followed by a burning sensation in the same place. In the skin, cold receptors outnumber heat receptors; however at extremes, heat and cold feel the same, registering as a burning pain. At 140 degrees, for example, both kinds of receptors are being activated together. This kind of pain can come also from internal chemical changes to internal organs.

Aching pain sensations travel at the same speed as burning pain sensations. But this kind of pain lasts longer and is more generalized.

All these acute pains are felt when the pain receptors sense chemicals created when tissue is damaged or note a deficiency in the normal oxygen-rich blood flow. Since sensory impulses from differing areas of the body sometimes have common pathways to the brain, the pain is sometimes felt in the "wrong" part of the body (a heart attack in the left arm, for example).

People seem to experience their pain quite differently, even though the intensity required to trigger the receptors does not vary much among human beings. Fear of a certain medical procedure can make its pain worse, as can fatigue and depression. Cultural differences in how you are supposed to react— everything from "holler" to "keep a stiff upper lip"—probably also can make it feel worse or better. Other things diminish your experience of pain; for example, excitement or intense preoccupation (these are probably keeping your nervous system pretty busy) and a sense that the pain is worthwhile (as in childbirth or athletic training). Knowing what to expect and being of advanced age cut into one's perception of pain too. The time of day when you are subjected to the pain can have something to do with it; more people can put up with more pain in the morning than at other times of day. Women also feel pain more readily than men (perhaps it is their oh-so-much-more-delicate skin?)

The body has another pain system, one for stopping pain. After acute pain, the brain and spinal cord neurons, along with the pituitary gland, release enkephalins and endorphins to make us feel better. They block pain impulses in a way similar to the effect of morphine. It is thought that what "exercise addicts" are addicted to is this release, which can be triggered by anything from acupuncture to skin irritants, even placebos, courtesy of the brain. The feeling of calmness after great suffering has been expressed beautifully by Emily Dickinson this way,

After great pain, a formal feeling comes—
The Nerves sit ceremonious, like Tombs—
The stiff Heart questions was it He, that bore,
And Yesterday, or Centuries before?

A little later in the poem, she refers to a "a Quartz contentment, like a stone."

Of course our pain and pain-blocking systems are hardly perfect. For example, because the brain has no pain receptors, brain tumors are not felt until they press against the skull.

And then there is chronic pain. This kind of pain, unlike acute pain, continues long past its proper warning time. It is caused either by emotional stress, a very serious physical condition, or both. Some 10 million Americans suffer from it; even if it is totally psychosomatic, in some cases it is also absolutely real. The brain's pain-relief system does not operate well against it.

Pain relief can be found through both chemical and surgical means. Some act by blocking the pain signal, others by blocking the reception of the signal in the brain. There are local and general anesthetics, which act on the pain signal and the pain reception, respectively; narcotics that act on the brain; and aspirin, which can be profoundly effective on the central nervous system. Other pain treatments include nerve surgery, which severs nerve pathways; electrical nerve stimulation, which blocks the signal; and hypnosis, which works best for organic illnesses or pains, by blocking perception of them in the brain. Biofeedback relaxation exercises and behavior modification also can be very effective.

Seeing Stars

Bam! Whap! And worse . . .

The "stars" you see when you hit your head are in no way astronomic but are flashing signals made by nerve cells in the retina, at the back of the eyeball. Ordinary seeing happens when retinal nerves fire and brain cells read, and all this usually happens when light hits the retina. Being hit on the head, by coincidence, causes the retina to conduct itself similarly. The misfiring—of just a few of our 100 million or so retinal nerve cells—is very brief and, I think, pretty. Nor is a blow to the head their only impetus. A small electric shock can also create the starry display, so can dizziness, even a moving magnetic field created in a lab with the help of powerful magnets.

A similar form of bodily "entertainment" courtesy of these retinal cells is the colored patterning that appears when you shut your eyes and rub them. But, lovely as these swirls are, didn't your mommy tell you not to rub your eyes? Or, certainly, not to hit yourself over the head?

Shivering

Shivering is the body's attempt to create heat. When chilled, your muscles begin to contract involuntarily at up to twenty times per second. As with any work, from a machine or living creature, waste heat is created as part of the equation. Presto—you feel warmer. This all works fine for a while, but it can ultimately drain a great deal of energy. Violent shivering and teeth chattering—the jaw muscles contributing to making warmth—are certainly not bad for you at all if they are short-term. But they are a signal to get indoors or otherwise get warm before you become exhausted and then ever colder.

It doesn't take much to get our bodies to shiver. Even a gentle wind can strip away the layer of warm air we all sport next to our skins, which is waste heat emitted just from the ordinary internal work of the "body machine." This insulating layer leaves us eight times faster in just a 5-mile-per-hour wind than it does if there is no wind at all. Being wet drains body heat yet faster, since water is a better conductor of heat than air is. Running outside in the rain on even a 50-degree day can make you shiver. And being in the water (from a boating accident or just swimming) drains body heat a full twenty to thirty times faster than being in the air. The wet of sweat is designed to cool us as it evaporates, so sweating in winter from wearing too many clothes during exercise can also stimulate shivering, sometimes too much of it. People with constantly cold hands and feet, perhaps from Raynaud's disease (a common condition), can also shiver more than others.

Since shivering can be either ameliorative or the first step toward freezing to death, it is well to consider this ultimate denouement. Once past the shivering stage—and the shivers will stop just as involuntarily as they began, even if you are still cold—frostbite can be the next step. Your exposed flesh may first feel slightly prickly or stiff, and it almost always looks a bit waxy. After this, the skin may begin to feel as though pebbles were being pressed into it. Later it can redden and swell, forming blisters and sores, or streak red in places. The last stage is when the body part—usually in the extremities—receives virtually no blood circulation and eventually simply blackens and drops off. If you think you might be getting frostbitten, use normal body warmth, as by tucking your hands in your armpits, and then get inside. Since frostbitten skin is very delicate, never rub it, or apply lots of heat, or treat it with snow or gasoline (these latter are old myths). The most severe cases should be quickly taken to the burn unit of a hospital, where skin grafts may be necessary.

Another after-shiver reaction to cold is, believe it or not, personality change. Memory loss—though just of the freezing incident itself—can begin as soon as your internal body temperature has dropped about 2 degrees. After that, people begin to act more peculiarly. Some who rarely swear begin to cuss a blue streak; others go into profound depressions. Most suffer a loss of judgment; boating accident victims who abandon ship to swim miles to shore, for example, are often exhibiting a reaction to the cold. So are the freezing victims who take off their clothes and run around.

More extreme physiological symptoms occur next if warmth is not available. Your muscles become rigid when your internal body temperature drops to around 90 degrees Fahrenheit. By about 86 degrees, the blood circulates so slowly that most people lose consciousness, and the electrolyte balance in their cells is severely disrupted. Death from heart failure occurs at an internal body temperature of about 71 degrees Fahrenheit in adults, though children can sometimes survive longer. The heart, which has entered fibrillation or uncoordinated twitching, simply cannot supply blood to the brain. This is not a shiver—this is the end.

Sleeping

To fall asleep is to place on hold, instantly, the most complex organ on earth, the human brain. When awake, our brains form more than 100,000 different chemical reactions every second, and these are the building blocks of every idea, feeling, and action we experience. In a single day, the average brain records almost 90 million separate bits of information. Why would we want it to go into "down time," rendering us both unconscious and immobile? Is sleep, as one researcher put it, "the biggest mistake the evolutionary process ever made"? It certainly seems strange that the evolution that has made for us so many protections in our waking state—everything from a trigger-sharp pain detection system to an exquisite sense of balance—allows all that protection to turn off, or at least turn dangerously imperfect, at night. And the sleep state adds up to a good chunk of our lives. Very curious.

There must be good reasons. But they have not been completely discovered. What happens and doesn't happen during sleep is a bit puzzling: The body does not seem to repair tissues during sleep, which rules out that as a reason for sleeping. We do not seem to need sleep to get us off our feet, since, even without the drag of gravity, astronauts want their sleep too. The pituitary gland does indeed release growth hormone during sleep, making this our best time to grow and providing one good rationale, at least for those of us who are still growing. The hormone prolactin, a chemical that is needed by nursing mothers and perhaps others, also peaks during sleep, and luteinizing hormone, needed to create the sex hormones, also surges. Researchers, coming at the mystery from the other direction, have also isolated a "sleep factor," a substance that is known to accumulate in sleep-deprived animals. It can build to as much as one millionth of a gram per 100 grams of brain tissue. What it is and why it is there are puzzling, but if this substance is injected into a wide-awake animal, it will quite quickly go to sleep. We are also more awkward at night, even when awake, but that might be an effect of sleeplessness rather than a cause of sleep. No other biochemical factors have yet been discovered to sort out this odd mix of findings.

It is known, however, that sleep is a time of complex brain activity. A look at our sleep cycles attests to that. In the first stage, light sleep, our blood pressure dips, breathing shallows, and body temperature falls. Brain waves adjust to sleep during this period of just 10 minutes or so. As they do, we occasionally jerk in a sudden muscular spasm called the "myoclonic jerk," thought to be caused by a sudden release of muscular tension engineered by a burst of brain waves, as they make the transition to sleep. The second stage of sleep is deeper. During this 20- to 30-minute period, the eyeballs begin to roll slowly behind the eyelids and brain waves begin to sharpen. The third stage is one of deepening relaxation, with temperature and blood pressure very low. And stage four is characterized by slow, rolling brain waves in their deepest sleep. Once this cycle has run, the first Rapid Eye Movement (REM) sleep period begins. This is the time for dreams (see the sections "Dreaming" and "Daydreaming"), and in it the brain is very active, the heart rate spikes up, breathing becomes irregular, and muscle tone is usually lost completely. These cycles then recycle throughout the night.

The reasons for sleep cannot proceed solely from brain complexity, though, or else why would animals sleep in such odd species patterns? People sleep fewer hours than most mammals, yet most mammals sleep more than other creatures. Lions and mountain gorillas, for example, sleep almost the entire day, and possums even longer (about nineteen hours in every twenty-four). Dogs have adapted to the human lifestyle pretty well but still conk out far more than people do. A few mammals, though, notably giraffes, take only "cat naps" while relaxing on their feet, as do most sea mammals and fish. There are a few fish that do not sleep at all, and neither do plants and insects, though they do seem to rest. Above the reptile level (evolutionarily speaking), most animals, with the possible exception of birds, also have the kind of brain waves during sleep that indicate dreaming in humans. A crazy-quilt, cobbled-together explanation here might have something to do with needing to save energy, but then resting is almost as good as sleeping in that respect. And we don't rest completely anyway: During sleep we move an average of forty to seventy times to keep blood flowing and oxygen balanced. Some explanations look for significance to the fact that predator species sleep longer than prey species, but in our early days on the planet we humans were most surely both.

The stages of sleep may provide clues to the reason for sleep. At least the heart and lungs get some rest in some of them. And perhaps we somehow need not only our dreams but the unconsciousness. The sleep habits of our early ancestors are not known, but gathering together at night is an ancient human custom. It has helped us get together to build solidarity—encouraging stable sexual and family ties—and has also gotten us out of harm's way in the dark night. Perhaps sleep is even designed to diminish the psychic terrors of the night. We are psychologically complex creatures indeed. Shakespeare put this beautifully when he wrote in *Macbeth*,

Sleep (that) knits up the ravell'd sleave of care,
The death of each day's life, sore labour's bath,

Balm of hurt minds, great nature's second course,
Chief nourisher in life's feast.

But then there are the puzzling differences in the amount of sleep required by normal people. It is said that Napoleon regularly slept just three to four hours a night, and Leonardo da Vinci and Thomas Edison napped for 15 to 30 minutes every so often around the clock, without additional night sleep. (Some of these stories may be exaggerated, of course.) It is normal for adult humans to sleep anywhere from four to ten hours a night. Short sleepers seem to concentrate on deep sleep and dreaming stages, though they do not equal long sleepers in these activities. The documented record for wakefulness is eleven straight 24-hour days, set by a young, healthy volunteer under medical supervision. He was scarcely productive or creative during this time, but he had no serious physical or mental problems. Certainly it didn't kill him to skip sleep. After the experiment, he slept fourteen hours and 40 minutes straight, woke up, and felt fine with no other catch-up required. It was once thought that sleep was not like a bank in which you could deposit, save, and withdraw, but it is now known that you can indeed store up a little sleep on the weekend for the week, and even when you go deep "into debt" you need not repay every hour borrowed in order to balance the books. As a species, we certainly do not need any set amount of sleep. Individuals vary, and each of us may well have our individual need that is genetically determined.

The sleep a person needs varies a lot according to where he or she is in the life cycle too. Newborn babies sleep sixteen to eighteen hours a day, then drop to about fourteen hours by the age of six months. This, coupled with the pituitary growth hormone release in sleep, suggests that growing may well be the most solid reason found so far for sleep (at least in children). Half of this baby-sleep is dream sleep too, which contributes to the argument that processing new information during sleep is important and may aid the maturation of the brain. Pre-schoolers sleep about ten hours a night, and adolescents about nine hours (the latter spend a great deal of time in deep sleep). Adults find seven to nine hours typical, while older people sleep four and a half to seven hours a night, less of it in dreaming sleep than they did when they were younger.

Sleep deprivation is common in the United States, sleep experts say, and they recommend more sleep for adults, to be achieved by going to bed earlier. Even with all the normal variation among people, you will know you need more sleep if your reaction time is down; this is obviously important when driving and operating machines such as cars and trucks, to say nothing of coming up with ideas at work and home. Though it hasn't been measured, researchers believe that enough sleep boosts higher mental functioning. And enough for you is probably not the same as enough for somebody else. It has proven to be impossible to train yourself to need less sleep, a regimen recommended to high achievers a few years ago.

Americans are chronic insomniacs, the strange other side of this coin. About 60 percent of insomnia is caused by mental or physical problems, sleep researchers think, and the other 40 percent simply by fretting over loss of sleep. And, contrary to popular belief, while alcohol and sleeping pills hasten sleep initially, there is a withdrawal about two to three hours after drinking alcohol that then wakes you up; pills, if they are antidepressants, can create an irregular heartbeat and urinary problems or worsen glaucoma, and they don't even work at all after a while. (There may be a role, though, for sedatives as sleeping pills, as long as they are taken on a very short-term basis.)

At least it is clear that sleep is a very sophisticated bodily function. When it goes seriously wrong, as in the excessive and inappropriate sleepiness called narcolepsy, it may be a sign of a severe neurological disorder or may proceed from other causes (a sufferer should visit a sleep disorders clinic for diagnosis). Then there's the story about the boy who chewed on his bedpost at night—sleep researchers finally decided he had multiple personality disorder and one personality took an animal form. There is clearly a lot more to be learned about our daily conk-out.

Sleepwalking

Sleepwalking may be the stuff of cheap, scary movies, but it is indeed real. And it is most common among children, especially boys, since younger nervous systems are not always mature enough to make the brain transitions required during sleep. In the case of sleepwalking, the sleeper has not passed smoothly out of the deep, slow-wave stage of sleep. About 15 percent of all children sleepwalk, though many do no more than sit up in bed and stare glassily into space for a few seconds, the bare beginnings of a stroll. Others may walk around for up to a half hour. No emotional or neural problems are indicated by this. Adults who sleepwalk might want to get a neurological and/or psychiatric evaluation, though adult sleepwalking is only rarely a sign of significant problems. It is also treatable.

Whatever the age of the sleepwalker, the old story that he or she should not be awakened is only partly true. A sudden waking might make the person confused and frightened and could conceivably encourage dangerous actions, since normal judgment is not operating in a sleeping person. There is no chance, though, that the sleepwalker will act out some odd or even drastic dream. (Sleepwalking does not occur during the dreaming stage of sleep.) The best advice is to guide the person back to bed without trying to wake him or her.

People who find themselves kicking their covers off hard, talking out loud, even jumping onto the floor while asleep are not sleepwalking but are indeed acting out part of a dream. Only a few of them may have REM Behavior Disorder, which rarely occurs often in any one person and most often occurs in older men. It happens when the muscular system, usually inactive during sleep to prevent us from enacting all our dreams every night, overrides that prohibition. Luckily this is very rare—it could, after all, involve everything from taking long drives to committing murders to seeking bed partners down the hall on a business trip, all while asleep.

Smiling

Edgar Allan Poe wrote in *The Purloined Letter* that, when he wanted to discover the true emotions of another person in conversation or from afar, "I fashion the expression of my face, as accurately as possible, in accordance with the expression of his, and then wait to see what thoughts or sentiments arise in my mind or heart." Actors know, too, how "making a face" literally affects your mood. So, as the songs say, "put on a happy face!"

You make those dazzling smiles with three muscles—the zygomatic (almost entirely), often the bicularis oris, with sometimes the risorius coming into play. Your mommy may have told you "A smile is easier to make than a frown" and, if she did, she was right. In fact, the smile is probably the simplest facial expression of all to make, muscularly speaking. Of course, sometimes you can actually smile so much that this muscle gets sore.

The grin probably came before the smile in evolution, though, since it is most likely an adaptation of the threat or fear expression found quite low on the evolutionary scale. A grin can be a mild type of threat in people too. Very few animals besides us, probably only the chimpanzees, can truly smile; but their "play face" is closer to what our smile means than their wide-lipped, "smile-y" look, since the latter means a more nervous or social excitement. And then there is the chimps' broad grimace, which indicates more intense fear or a general high excitement.

Of course, not all of our own smiles are signs of simple joy either. There are fake smiles, bitter grimaces, ingratiating smiles, smiles of embarrassment, of fear, of contempt, of compliance, of flirtation, as well as those designed to soften the making of a negative comment. There are even smiles of misery. One researcher claims to have differentiated fifty different smiles. Winston Churchill said of the Soviet leader Molotov, "(He) has a smile like the Siberian winter." And Shakespeare has Othello remark, "One may smile, and smile, and be a villain." We have evolved to be subtle creatures indeed.

The biggest difference of all in the smile realm is said to be between the genuinely felt and the fake smile. These two smiles are even connected differently to the brain: The fake smile is created by nervous system passageways to the evolutionally newer cerebral cortex. And the truly felt smile is neurally hooked to the more primitive and older limbic system of the brain.

These two smiles of ours differ in their musculature too. In the truly felt smile, the "zygomatic major" muscle does almost all the work—it lifts your cheeks up, pull up the corners of your mouth, stretches your lips, and slightly bags the skin below your eyes. At the same time, the bicularis oris muscle around the eye makes your eyes narrow a little and the skin near them crinkle. The false smile, on the other hand, is slightly asymmetrical and, more important, it almost never crinkles the skin around your eyes or flattens them slightly. The true smile is also somehow "contagious"—we smile when others do, often even if they are actors in plays or in commercials. And the people involved with plays and ads know this too.

Some of our other smiles use the muscles slightly differently. The fearful smile, for example, adds in the risorius muscle that makes the mouth into more of a rectangle. The contempt smile tightens the muscle in the corner of the lips, to make a slight bulging there. The flirtatious smile, like the one in which Mona Lisa is caught, also involves a quick look by the eyes to and away from its subject. The smile that we use to soften a difficult emotion (such as when a boss smiles after criticizing an employee) often adds a head nod. And so on. It is very rare, though not impossible, for any smile besides the true one to involve the characteristic crinkling and narrowing of the eyes.

Sneezing

          A sneeze is a particularly violent breath. If you have a typical cold, your sneeze will propel about 2,000 virus particles from your nose at the rate of 75 to 100 miles per hour. If your sneeze is related to an allergy, you will ah-choo out a blizzard of histamines (body by-products) inside your nose and then out farther, toward all of us.

          Let's first take the sneeze from the common cold. This is such an innocuous name for something that can make you feel so nasty. According to the American Medical Association, the common cold has a $5 billion impact on our economy because of doctor visits, lost days at work, and sales of patent medicines. The horridness of the cold often begins in sneezes, created by one of the 200-some different cold viruses afoot in the world. (No wonder it has been hard to come up with a vaccine, and no wonder that you can get a lot of colds—almost all of them only once—in a lifetime.) The cold virus you get commandeers the genetic machinery of the cells in your respiratory tract, tricking them into manufacturing more cold virus. So it gets worse. But then, by the end of a week or so, the body has not only wised up to this but has fought off the virus. At that point, your last sneezes will be composed, in the nose, almost entirely of quite dead virus particles.

          At the height of your cold, though, the sneeze you sneeze onto the desk, coffee table, kitchen counter, door handle, or railing is very much "alive." And the virus particles in it can continue to live for about three hours on the sneezed-upon surface. Anyone who puts a hand down on the surface, then touches nose, mouth or eyes, can pick up your cold.

So cover your mouth, for cripe's sake. Not only that, wash the hands you used to cover your sneezing mouth and nose. Even just rinsing them in plain water for 30 seconds will wash the bits of virus down the drain. Otherwise your next handshake or hug could be very un-nice. Picking up a cold from touching something is much more common than getting it from being downwind of a sneeze itself. A large enough dose of virus particles, though—no matter where they come from—will give the beginning of one to anybody, anytime, unless that person has already had that particular cold.

The period during which a cold sufferer sheds the largest numbers of virus particles, and so is the most contagious, lasts longer for children. They shed for two to four days, versus one to two days for adults. Children's immune systems haven't had enough practice in manufacturing natural interferon or at recognizing infections to gear up any faster. And, of course, they can forget to cover their mouths, then "bless" you with their sneezes.

The upshot here is that you get colds from other people's virus particles, not from "getting a chill" or something. People get more colds in winter for three simple reasons: They are inside more with other people, some of whom are bound to have colds; the "rhino virus" family that causes most of our colds is more active in winter; and our nasal membranes can become dry and cracked from low-humidity air, allowing a good foothold ("nosehold"?) for the virus particles. If you still don't believe me, ponder Antarctica. Researchers there get new colds, but only when new people join the group; otherwise they are cold-free in a pronouncedly cold place.

None of the usual nostrums really cures the cold and all that comes with it. Taking antihistamines dries your nose, yes, but it can thicken the mucus in your lungs, encouraging bronchitis. Antibiotics don't combat viruses, only secondary bacterial infections. And those over-the-counter drugs can interfere with a medicine you are already taking, as well as make you drowsy.

"Aspirin and lots of fluids" is still the best advice. The aspirin works on the inflammation and the fluids actually dilute the virus while helping the body drain it. Every sneeze gets rid of some of the virus too, and, contrary to the old stories, does not let your soul out and the devil in.

Sneezing from allergies is different, and just as uncomfortable. Here, your nose is trying to do you a favor by getting rid of the irritant (pollen, dust, mold, dog or cat dandruff, even cockroach feces, etc.). Inside your nose's mucous membranes and your whole respiratory tract are cells called mast cells, which go on alert at the entrance of the allergen. Onto these mast cells then binds a substance called immunoglobulin E. At the quick second whiff of the allergen, the whole little system explodes. This releases the histamines that then tell your brain stem to—lickety-spit—order a reflex action to get rid of it all. That action is the sneeze. (Other reactions produce the stuffy nose and slippery mucus.)

There are always new drugs for asthma and other allergies, including Intal/Nasalcrom/Opticrom, derived from cromolyn sodium (ask your doctor). And there are also "sneeze calendars" to tell you which parts of the United States have which pollens in which seasons, and so on. But the main thing to remember is that the sneeze, at least, is trying to help you (even though it makes you close your eyes and maybe get into a car wreck or something). Don't hold it in—that's hard on your eustachian tubes and eardrums. Cover up, and sneeze away.

Snoring

The story goes that George Washington, Abraham Lincoln, and Benito Mussolini were among the world's stellar snorers. But, chronic snorers aside (far aside), everyone snores sometimes, including children. Even you, dear reader, will buzzsaw at least an occasional night away.

Snoring comes from blocked air passage to the lungs. The lungs must then draw air in hard to compensate. This vibrates the uvula (the flap of soft tissue that hangs down in the back of your mouth) as well as the roof of the mouth itself. Nasty noise is the result. There are many common causes for common, once-in-a-while snoring. If you have a cold or respiratory allergies, you will breathe poorly through your nose at night, your mouth will open wide to take over, and it still will not access your congested airways well. If you overdrink or oversmoke or overeat one evening, your airways may become swollen or obstructed too. Temporarily enlarged tonsils or adenoids, dentures that don't fit (or are removed for the night), and sleeping too much on your back are also garden-variety causes of snoring that can affect anyone. And, as for noise, these ordinary snores have been clocked as high as about 70 decibels, about like a jackhammer at a few hundred feet away.

Chronic snorers—about one person in eight—have been measured blasting away louder, at nearer to 80 decibels, as though the jackhammer were hammering at just 10 feet away. And their average tally is a bit more than 1,000 snores per night per person. There are usually different causes here. Habitual snorers are either obese, have a deformity in their nose or a

chronic enlargement of tonsils or adenoids, or always overdo
in the alcohol or salty food department in the evenings and
sleep on their backs most of the night.

There is one even more serious cause of constant snoring:
sleep apnea. In this condition, the sleeper actually stops
breathing for a few seconds, even a few minutes, at a time,
and this happens several times or more every hour. Though
this condition can be fatal, most sufferers wake up automatically and suddenly gulp air. As they gulp, they snore. They
also wake up unrefreshed in the morning, since their sleep has
been riddled with these tiny forgotten awakenings. Not every
chronic snorer has an early stage of sleep apnea, but it wouldn't
be a bad idea to visit a doctor and find out if you or a sleeping
partner does. To stop the snoring, yes, but also to correct this
condition that deprives the person of a full measure of oxygen
in a lifetime. Many doctors even think it contributes to both
abnormal heart rhythms and high blood pressure as the heart
must beat harder, every night, to provide more air. About 90
percent of sleep apnea sufferers are obese men, and they can
recover when they lose weight or have any back-of-the-throat
blockage removed by surgery.

There are many ways to treat common or chronic snoring,
some good and some bad. First the bad methods: Never use
any device that keeps the mouth closed (you need air, no
matter what) or one that prevents you from turning side to
side. Better ideas: Take antihistamines or decongestants for
allergies and colds that lead to your excess sonorousness, but
avoid them within two hours of going to sleep; and stop smoking or using alcohol and tranquilizers. You can also sleep in
one of those collars used by people with neck injuries, or prop
the top of your bed up with bricks on those two corners, or
sew a marble into the waist on the back of your pajamas (to
discourage sleeping on your back). Higher-tech solutions include a mask and a bedside tank that shoves oxygen into your
nose, as well as surgery to remove your uvula, deviated sep-

tum, or other relevant physical problem. The U.S. Patent Office has about 300 "snore cures" on file, including one little music box to attach to the back of your pajamas—it warbles, "Roll over, dear" when you hit it.

Of course, none of this is a joke. Snorers and those who sleep with them can suffer greatly from sleep deprivation. This can cause headaches in the mornings, defects in memory, also impotence and depression. To say nothing of fury at the snorer.

Spitting

The average person, not even the obvious slobberer and drooler, secretes more than 2 1/2 pints of saliva every twenty-four hours. This is not designed for depositing on the sidewalk, but for starting the digestion of your food and for keeping your mouth moist enough to talk, whistle, sing, kiss, and so on. So be grateful for this spit!

The saliva gland is under your tongue. What it emits is a slurry of mostly water but also mucus and the enzymes amylase and maltase. The latter pair start your digestive process by breaking up starches into a simpler sugar. The whole slobber sop moistens the food so that it can get down the esophagus.

We secrete saliva at a greater rate, about twenty times faster, when we expect food in the mouth, but a little is coming out all the time. And, since the nose and mouth are connected, viral infections like colds and postnasal drip add live and dead virus cells, as well as more mucus, to what's in your mouth. So there is variety to your spit.

Taste buds are mostly on the top of the tongue, but they couldn't taste anything without the moistening of the spit.

A horse secretes about 9 gallons of saliva in twenty-four hours, and a cow about 12 gallons. Think we humans are on the dry side? Well, add ours up: Over your entire life, you will probably produce enough spit to fill a generous-sized swimming pool!

Stomach Rumbling

It can be embarrassing or just mildly annoying, depending upon where you are and who you are with. And it can occur before and even after your meal. The rumble before becomes more of a gurgle afterward, and they are two slightly different phenomena.

Whether you are being introduced at a party, or chairing a meeting, or snuggling with a favorite person, your stomach is going about its business. Every 75 to 115 minutes its muscles contract. When no food is present, their rhythm is a wave-like stretching and contracting that molds the air, mostly digestive gases, in the stomach cavity. No one quite understands exactly how this makes the tummy rumble noise, but it surely does.

After dinner, the noise is sloshier, of course; it is created when gas bubbles are trapped in the slurry of food as stomach contractions reach their active peak once every twenty seconds or so.

Sweating

"What dreadful hot weather we have! It keeps me in a continual state of inelegance," Jane Austen wrote in a letter to her sister. Her careful reticence, if not her style, is still sounded in the old saw, "Horses sweat, men perspire, and women glow." But let's call it even more than moisture, from the 2.3 million sweat glands that cover the average person—let's call it sweat.

Not all sweat smells, since most of our sweat glands are eccrine glands. These secrete water, salt, and mere traces of chemicals such as copper, iron, mercury, lactic acid, and urea. The latter is the chemical found in its highest concentrations in urine, which is why you don't need to urinate as much on a hot, sweaty day. The more physically fit you are, the more you sweat in this way since your body has learned well how to shed heat. This eccrine brew usually stinks only when there is an ample bacterial buildup on the skin to perfume the fluid, a buildup that often comes from the lack of a bath. Foods can scent our basic sweat too; garlic, alcohol, and chocolate are common culprits.

The rest of the body's sweat glands are aprocine, concentrated mostly in the armpits and pubic area. They are designed to do their thing only during emotional stress and become active only once you reach puberty. The scent they create is designed to be sexually exciting; if you don't like your partner's odor, switch partners. (There is more, maybe more than you want to know, on this and related matters in the section called "Stinking.")

The third kind of sweat, called insensible perspiration, is

the most basic and the least noticeable. It is really just the airing of the body through the sweat glands and skin pores, but to the tune of nearly a quart a day of liquid whether you exert yourself or not. It is released so gradually and evaporates so instantly that it does not "puddle" on the skin.

I hope it is this latter sweat that Thomas Edison meant when he said, "Genius is one percent inspiration and 99 percent perspiration." But whether you are a genius or not, "don't sweat it," just work "by the sweat of your brow," and "never sweat the small stuff" or "let them see you sweat." But I digress.

The basic reason for all of our sweating is to get rid of excess heat and to cool the body at the same time. Like any machine, our bodies use energy to power internal functions and keep us moving around. And, with any kind of energy, even from a single light bulb, there is waste heat. If our own waste heat were not emitted, our temperatures would soon rise so high that we would die. In the case of all mammals, this heat exits the body through sweat.

And here is how it helps to cool us at the same time: The little droplets of sweat evaporate in the air, using up heat in making this phase transition from a liquid to a gas (in this case, becoming part of the gas called the air). At the same time, the blood vessels near the skin's surface dilate, allowing for greater blood flow there and thus more radiation of heat away from the body. On very humid days, this cooling evaporation does not occur well since the air is already so saturated with water—that's what makes it harder for us to cool off.

While you may be more worried about embarrassingly sticky clothes and damp shoes on days like these, your brain also has to get rid of excess heat. This all-important organ heats up because the blood does, and this hotter blood could even start to poach your brain. Fortunately, we have evolved a very neat trick in this department. The blood in your veins— which has not been heated like the blood in your arteries, because it has not rushed to help you move around—actually changes direction. In two veins in your head, this cooler venous blood goes back into the brain to cool it. Of course if you are stupid enough to exercise hard on very hot days, you can override this protection—and poach.

This heated scenario brings me to what happens beyond normal sweating: the four stages of heat reaction up to death. The first is sweating to the point of salt and water deficiency, which can actually reduce the volume of your blood since blood is mostly water. In this condition, less blood is available to crucial internal organs such as the heart, kidneys, and brain. If you approach this point you will know it by the dizziness and the feeling of faintness. The answer is not to scarf down potato chips before you run in summer but to avoid exercise outside when it is terribly hot. And even on normal summer days, it is good to "preload" with two cups of water, then drink more every fifteen minutes. Do not rely on your thirst to guide you here since you can sweat out water at the rate of two to three quarts per hour during exercise in heat, too fast for the thirst signals to react well enough. This latter should be telling you something too.

The second stage of serious heat reaction is heat cramps. These are muscle spasms that proceed from an electrolyte imbalance in your cells caused by loss of water. This means stop and get cooler. These cramps can also be forestalled by gorging on water before moving around a lot in the heat.

True heat exhaustion, next, is even more dramatic. It involves fainting, profuse sweating, and then proceeds toward no sweating at all, which is particularly dangerous. Lie down immediately in a cooler place, head down and feet up.

The last stage is heat stroke. This is the end of all sweat. The dizziness here is worse and comes with plenty of abdominal pain, confusion, and, as internal body temperature spikes up, delirium, collapse, seizures, and unconsciousness. All show that death is near. Get cold water onto the victim fast and pack ice around him or her. Then go straight to the hospital—brain function is in danger and so is life.

I guess I hate to end this way. So how about a few fun facts on sweating: You can't "sweat like a pig" because pigs sweat only from their feet bottoms and their lips. People who seek a doctor's help for chronic heavy sweating are often zapped in some of their sweat glands with low volts of electricity, to close them up for several weeks. Most antiperspirants close sweat glands with aluminum salts. As you get old, you sweat less. Emphatically spicy foods make you sweat, which is probably why they are so popular in hot climates. And the mammary glands, for milk, are just modified sweat glands.

Sweaty Palms

If someone stretched out the skin of your whole body—God forbid—there would be 17 to 20 square feet of it. And on every single square inch of this skin there are 625 to 650 sweat glands. The sweat glands on the palms of the hand and the soles of the feet are slightly different, because our skin is thickest there, but they are no less numerous. Some days, it may seem as though they are all draining water (our adult bodies are two-thirds water anyway).

Sweat is designed to cool us (see the sections "Sweating" and "Stinking"), and nervousness or stress makes us warm enough to need this helpful action. Under stress, the hypothalamus orders the release of adrenalin and noradrenaline to gear us up for emergency action, and part of this bodily buildup is a signal to hold heat in the internal organs for mobilization of heart and lungs, but to let it go out of places such as the palms, soles, and crotch. So if you get sweaty palms, you also have, at the same time, sweaty soles and a sweaty crotch. (But at least no one has to shake hands with them if they don't want to.)

People with not occasionally but chronically sweaty palms have the condition known as hyperhydrosis. They could decide to use an antiperspirant on their palms, learn biofeedback, have water forced into these sweat glands electrically until they won't work, or even have an operation that cuts the nerves that control those particular glands. If you just get it occasionally, though, don't worry—everybody does.

Tickling (Yourself)

A major mystery is why you can't tickle yourself. It makes for a major goal—understanding tickling. And there are two kinds.

The first is the gentle tickle, a light touch to the skin. This works especially well as a tickle, making us titter, when the area of skin that is touched is one where "free nerve endings" are densely packed, under the arms, for example. These nerves, actually slower to communicate sensation than our specialized nerve endings but fast enough, send a message to the brain at about 3 feet per second. The skin—our body's largest organ—is on us, after all, to tell our brains about the outside world and whatever is coming at us from it. It might be the brush of a big branch of a tree on our path, or a mosquito, or a snake starting to slither onto our feet. We have evolved to react to these light touches with a quite primitive and sturdy reflex of "uh-oh, what's that?" When the potential danger proves harmless, we laugh and it is a little nervous laugh of relief.

The second kind of tickle is the gigantic, massive tickle. This level of "assault," more within the domain of the sympathetic nervous system, is a major threat that creates tremendous relief when it is (very quickly) found harmless. We laugh hard. If the tickler continues, though, it can actually come to hurt. This is the body warning you that it is too much. This is why violent tickling is cruel.

In neither case—the light intrusion or the rough touch—would it make sense to react to our own actions in the same way. We simply can't go on alert or find any ancient danger in what we are doing to ourselves. You "can't fool mother nature" in the form of your free nerve endings or sympathetic nervous system this much. And if there is no danger, there will be no laugh, giggle, or titter at its end.

Trembling

We tremble from emotions like fear, happiness, guilt, and excitement—even from simple fatigue. Our bodies are usually under our control, but trembles, like cramps, twitches, spasms and temporary paralysis, are important neural-muscular exceptions.

The tremble is a delicate artifact of muscles and nervous system. Our nervous system is a network of more than 20 billion cells communicating through electrochemical impulses and masterminded by the brain and spinal cord. Much of this incessant action is conducted below the level of our consciousness; otherwise we would get nothing else done. There are many different types of nerves in the system, and motor nerves are one division of the labor. They tell the muscles where to go, what to do, and how fast to do it. After all, athletes are training their motor nerves as well as their muscles. With all the complexity involved, minor things can often go wrong with the motor nerves. (Major things would cause spasms.) And, when you add in small aberrations in muscle contractions within this dynamic duo of muscles and nerves, it is easy to see how trembles can shake you for a while. The nonemotional trembling caused by fatigue can be blamed mostly on the muscles. As for trembles from emotion, look mostly to the motor nerves, though they both always work together. Usually it's a bit less calcium outside these nerve cells that makes them send tremble messages, and our calcium balance is affected by the chemistry of emotions.

Trembling is usually very temporary and nothing to worry about. Your nerves and muscles will get their act together again pretty quickly.

Urinating

Your average daily output of urine is 1 1/2 to 2 quarts. (Aren't you glad to know?) Urine passes from your bladder through two valves, one of which opens automatically, and one of which, closer to the outside, is under your control (we hope). You can hold about 1 pint without exceeding your bladder's maximum muscular capacity. And signals are provided by the bladder as this amount is approached.

What is urine composed of? It is the stuff left over from your digestion of protein, plus uric acid, creatinine, urea, and small amounts of other material from the gall bladder.

To make a daily allotment of this product, your kidneys must filter about 48 gallons of blood. This is because blood is where your food and everything else is suspended. They are busy little organs indeed.

Your kidneys work harder under certain circumstances to make their favorite product. If you are afraid or very anxious, your blood pressure will rise and busy the kidneys to make more urine. When you are cold, the blood supply to the skin is lowered, making more for your kidneys to work on in their oh-so-skillful way. Caffeine, alcohol, and other diuretics stimulate the kidneys too. The digestion of a meal gets your whole body, including the kidneys, going. And, of course, pregnancy is all about a little baby sitting right on top of your bladder, and kicking every time it feels like it.

What cuts into your urine supply? Being very hot, which encourages lots of liquid to exit through the skin, smoking cigarettes, and probably, going to bed for a night of less activity than in daytime.

The urinary tract develops problems occasionally, more often in women than in men because the female urethra, the tube that connects the bladder to the outside, is much shorter in women and so can let more germs in. Sometimes urinary tract infections are caused by too many bacteria around the opening of the urethra; they then get inside the body. Some people even have an inherited cellular abnormality in the urethra's lining that makes it hard for the cells to fight off this bacteria. Others have urine that allows more bacteria to grow. Both groups will have many more of these infections. Some, but not many, women can get chronic urinary infections from sex, especially from a shoving against the opening of the urethra, or from a diaphragm that is too large and so pushes against the neck of the bladder. Drinking plenty of water before sex and urinating right afterward help in both cases. In general, bacteria can lodge and multiply in the urethra, the bladder, or the kidneys, causing a different disease in each place. The worst urinary tract infection of all is interstitial cystitis, a terribly painful chronic inflammation of the bladder; its cause has not yet been fully determined. If you suspect you have any of these infections, see your doctor.

Vomiting

The stomach is shaped like a boxing glove and, when you're ready to vomit, you may feel like the victim of a gut punch. But you should be glad—vomiting is the gastrointestinal system's equivalent of coughing, and it is trying to help you get rid of something, or at least to slow you down while you regain your health. The reasons for vomiting include food poisoning (usually from some salmonella bacteria, staphylococcus bacteria, or botulism bacteria), indigestion, tonsillitis, an obstruction, a variety of infections, motion sickness, smells, brain tumors, meningitis, intestinal parasites that cause diseases such as amebic dysentery, pregnancy, and more.

It is not really your stomach that conducts the crucial motions to make you vomit, though. First the diaphragm contracts and drops down; then the abdominal wall muscles become rigid. These operations raise the pressure inside the abdominal cavity and squeeze the stomach, which lies between diaphragm and abdominal wall. Finally, up the stuff comes through a convenient opening for equalizing pressure—your mouth. While this is going on, the larynx rises so that the vomit doesn't enter your airways; since the larynx can't move when you are unconscious, vomiting is dangerous then. For some unknown reason, your eyes water when you vomit too. Because the diaphragm is the main actor in vomiting, it could be said that it is a form of breathing or coughing.

Now for a really strange fact: Herbivorous animals (like horses and deer), as well as rodents, almost never vomit. If a horse ever does vomit, it comes out of its nostrils instead of its mouth. I can't say anyone knows why this is true.

But let's end on a classier note. Shakespeare wrote in *King Henry IV*, "How has he the leisure to be sick/In such a justling time?" You probably don't have the leisure time to be vomiting either—so I hope it happens very rarely.

Whispering

Talking uses your vocal cords, and the faster and harder the air is exhaled, the louder your voice will sound. Whispering doesn't use the vocal cords at all. In it, you are pushing a soft blast of air over tissue folds in your throat instead. These folds are in the larynx too, but just a bit above the vocal cords. The air then bypasses the nose, mouth, and other amplifiers. Pssst. Can you hear me?

Wrinkling

You don't have to be drastically old to notice your first wrinkle. And then they multiply, showing all the world whether you have smiled more than you have frowned or worried, crinkled your nose up more than you set your mouth in a sullen grump or grimace or grin. Wrinkles are character lines, my mother says.

Here is how it happens. The elastic fibers in the skin both thicken and snap as you age and/or assault your skin. The thickening is what makes leathery skin, and the snapping loosens the skin, allowing it to droop and fall into the tried and (thereby) true ways of your facial muscles. Since the skin around the eyes is thinner, it falls into wrinkles most obviously. More dead cells also litter your skin's surface as you age too, since fresh cells are replacing them at a slower rate.

Wrinkling is vastly enhanced by exposure to the sun and by smoking. On a more temporary basis, salty food at dinner will make liquid actually pool under your eyes at night; by morning, there are those puffy little pouches outlined by lines. As your face loses elasticity, fluid retention like this becomes more common too. Loss of sleep can make you look more wrinkly also, since the facial muscles that hold everything up are saggy. But you are bound to have some lines anyway. Don't sweat it.

Yawning

           People used to think that an unguarded yawn was dangerous, since it provided a path by which the spirit could escape the body. Actually, yawning is healthful. It is a reflex originating in the brain stem by which a deep breath is taken. The goal seems to be either to get more air to the lungs, and thereby the blood, and everywhere it goes, or to equalize pressure in the ear. In the first case, the yawn is triggered by a buildup of carbon dioxide in the blood, a signal that the oxygen balance needs to be redressed. This buildup occurs most often when you are breathing slowly or shallowly, a common situation when you are tired, sitting still, or confined to a hot, stuffy room. Inhaling as one yawns brings in oxygen, and exhaling releases carbon dioxide. The extra oxygen makes the yawner feel a little more alert. In the second case (the one that involves the ear), the yawn is triggered by a difference in pressure building up on one side of the eardrum. To try to equalize this pressure, you swallow or yawn, opening up the eustachian tube, which is near the middle ear, to let air in. (See also "Ears Popping.")

    Yawning is also curiously contagious. Unless you make a conscious effort, you will yawn when other people do. The spasm of muscles in mouth and throat will throw your mouth open, perhaps embarrassingly. Though we often feel bored or tired when others do, and probably are experiencing their need for extra oxygen at the same time, no one completely understands why yawning is this "sympathetic."

    One more thing is known, however: We are not alone

in our yawns. Species from birds to hippos and rats to fish also yawn . . . probably to wake up a little and get going, possibly to synchronize with other species members yawning then too.

Afterword

Well, I hope you have had fun reading all about your functions. Do you think some of them sound a little yucky? Maybe you don't fancy snoring, stinking, blowing your nose, farting, choking and so on? But be glad you can do everything in this book because, even though other animals can do a lot of these things, maybe even 90 percent of them or so in the case of the great apes, humans are the only ones who can do absolutely all. Have you ever seen a cat sleepwalk? Or a dog cry? Or even a chimp daydream—for sure? I think we're wonderful. In fact, I think we're a miracle of evolution. Maybe you do too, especially now, after reading this book.

If you still don't feel appreciative of dandruff, burping, and hitting your funny bone, contemplate as an alternative that ultimate bodily function: dying. Let me tell you about it! Your skin quickly loses its elastic quality and color, and the muscles their tone. Blood pools in the parts lowest to the ground, making parts of the body look bruised. Then you start to cool, at an average of about two and a half degrees per hour. You are stone cold after about twelve hours, and fully stiff with rigor mortis before that. The next step is for you to turn "flopsy" again. And decomposition, conducted by your own bacteria, begins to be noticeable after forty-eight hours at normal temperatures.

Now you should be delighted at coughing, seeing stars, hiccoughing, even itching. And certainly undyingly grateful for smiling and laughing and dreaming.

Bibliography

Brody, Jane. *Jane Brody's Guide to Personal Health*. New York: Avon Books, 1982.

Brunn, Ruth Dowling, and Bertel Brunn. *The Human Body*. New York: Random House, 1982.

Bryan, J., III. *Hodgepodge Two*. New York: Atheneum, 1989.

Diagram Group. *The Human Body*. New York: Facts on File, 1980.

Ekman, Paul. *Telling Lies*. New York, London: W. W. Norton, 1985.

Farndon, John. *The All Color Book of the Body*. New York: Arco Publishing, 1985.

McAleer, Neil. *The Body Almanac*. Garden City, New York: Doubleday, 1985.

Miller, Jonathan. *The Body in Question*. New York: Random House, 1978.

National Geographic Society. *The Incredible Machine*. Washington, D.C.: National Geographic Society, 1986.

Nilsson, Lennart. *Behold Man*. Boston, Toronto: Little, Brown, 1974.

Powis, Raymond, C. *The Human Body and Why It Works*. Englewood Cliffs, New Jersey: Prentice-Hall, 1985.

Prevention Magazine. *Future Youth: How to Reverse the Aging Process*. Emmaus, Pennsylvania: Rodale Press, 1987.

Smith, Anthony. *The Body*. New York: Viking, 1986.

Time-Life Books. *Mysteries of the Human Body*. Alexandria, Virginia: Time-Life Books, 1991.

Wallechinsky, David, Irving Wallace, and Amy Wallace. *The Book of Lists*. New York: William Morrow, 1977.

Wallechinsky, David, and Irving Wallace. *The People's Almanac*. Garden City, New York: Doubleday, 1975.

—*The People's Almanac 2*. New York: Bantam Books, 1978.

A Request for Letters

Readers who have had fun and been inspired by the information here may have questions to ask the author, or perhaps may want to suggest other bodily functions they'd like to hear about. If the response is overwhelming enough, I'll do a sequel!

Send your letters to:

Editor of *Why Can't You Tickle Yourself?*
Warner Books
1271 Avenue of the Americas, 9th floor
New York, NY 10020